POULTRY SCIENCE

The Selection, Rearing and General Treatment of Poultry

BUFF COCHIN COCK
Winner of 1st and Cup, Crystal Palace.
(Property of the late J. Partington, Esq.)

The Poultry Science

The Selection, Rearing and General Treatment of Poultry

L.C.R Norris-Elye

BIOTECH
2024
Biotech Books

ISBN13 978-81-7622-144-3

Published by : **BIOTECH BOOKS**
4762-63/23, Ansari Road, Darya Ganj,
New Delhi - 110002
Phones : 011 4300 3222
Email : biotechbooks@yahoo.co.in

PRINTED IN INDIA

Contents

Chapter 1
Introduction

Though the number of persons who keep poultry is annually increasing, and the knowledge of how to manage poultry is being diffused through the Poultry Press, the demand for eggs and dead fowls increases in still greater degree. This is principally owing to the increase of the population of these islands; and also to the better standard of living which extends through all classes down to the poorest cottager, and the consequent demand for eggs--no longer as a luxury, but, at any rate in some months of the year, as a cheap and nourishing food. The demand is to some extent met by the spare birds and eggs of the thousands of fanciers spread all over the country, who, breeding for show purposes, have a good many so-called "wasters" available for farmyard or eating purposes, and after they have hatched enough chickens, have also a good many spare eggs which, if they do not pickle, they are glad to sell.

The fact that some five thousand millions of eggs are annually consumed in the British Islands is sufficient evidence of the importance of the interests involved in the poultry question; that no less than some two thousand millions of these have to be imported from abroad suggests that by some means the money expended in these purchased eggs should go into the pockets of the British farmer and cottager. The latter, especially when he lives in the country and in a country lane, has no doubt great advantages for making a good addition to his income by keeping a few good layers, which can be attended to by his wife and children, and can find a large proportion of their living on the roadside and in neighbouring grass fields, over which he may be able to get permission for them to roam, to the great advantage of the fields in question; for they will keep down grubs, etc., and enrich the land by the very valuable manure.

Nearly every breed, nowadays, has a club of its own, managed by the leading fanciers of the breed in question, and from time to time settling the points of the birds. At the head of these clubs stands the well-known Poultry Club, with its membership of close on 1,000; the other clubs are mostly associated or in alliance with the Poultry Club, and send representatives to sit as members of its committee. Before this committee are brought various questions for decision, especially such questions as involve charges of fraud in exhibiting. These matters are of interest principally to exhibitors, who, though they number many thousands, are very few in number as compared with the hundreds of thousands who keep poultry for egg and table purposes. But from these latter the ranks of exhibitors are constantly being recruited, as they make the discovery that it costs no more to keep and feed a good fowl, worth from 10*s.* 6*d.* to £50, than to keep a fowl worth, at its best, only 2*s.* 6*d.* to 3*s.*

The writer himself began with a few fowls for laying purposes, but the Show records of the last twenty years go to prove that in his case, as in so many others, the ideas enlarged with the opportunity.

To a beginner I would say–Consider your object carefully. Do you aim at eggs only then select your birds for that purpose. If at table poultry, a different breed should be chosen. If to combine both purposes, some breeds may be found which will well answer both objects.

Domestic poultry are no new invention: Jews, Egyptians, and Greeks kept them, and the latter used them for divination purposes. The Romans, who took so much else from the Greeks, as the Greeks themselves had previously borrowed from the Egyptians, no doubt derived from them the idea of ascertaining the future by observing the feeding of fowls. The story of the disgusted Roman admiral, who finding that he could learn nothing in this way from certain fowls which refused to eat when wanted, flung them overboard, with the exclamation that if they would not eat they should drink, is well known. That the Capitol of Rome was *saved* by the cackling of geese may be true, though the experience of later days seems to run in the opposite direction. At any rate, so far as England is concerned, it is believed that the Dorking has been handed down to us from Roman times, and may possibly have been introduced by them. The Game fowl, also, is of great antiquity, and has been highly prized in England for centuries. Of late years many new breeds–some of which ought not strictly to be called *breeds* at all, but are carefully selected cross-breds from other fixed breeds–have met with some favour, and, at their hest, are handsome as well as useful fowls. Many of these, however, have not yet been sufficiently fixed to breed true with certainty, though this difficulty will no doubt disappear as time goes on. The most striking claim as to origin was that made some years ago, when it was stated that the Langshan was allied to the wild turkey! This idea is not broached nowadays, though the Langshan, like the wild turkey, is undoubtedly good to eat. The origin of many Bantams is to be found in the great interest taken by the Chinese in dwarfing each breed: for instance, the Pekin Bantam, the first known specimens of which were found in the Emperor's Summer Palace, near Pekin, when captured by the British and French troops in the China War of 1860. Many other breeds of Bantams have been later developed by the skill of English breeders, especially by the late Mr. Entwistle. There seems to be a present rage for novelties of a buff colour, in the manufacture of which, no doubt, buff Cochins have been of use. These new buff varieties have, however, at present great drawbacks, if it is wished to consider them as pure breeds, which they certainly are not. They at present will not and cannot be expected to breed true, being new compounds of other breeds: the Dorking and Buff Cochin having more share in the matter than other breeds. Moreover, the new buff breeds have a fatal family likeness, and it would probably be possible to obtain Buff Rocks, Buff Leghorns, Buff Orpingtons, and Lincolnshire Buffs from a single breeding pen, the same sire being used, but four slightly different hens. That this craze for newly compounded buff birds has been of great injury to the fancy at large, and led to many more valuable breeds being for the time neglected, there can be no doubt.

Fortunately, with this desire for novelty, there has, for the last year or two, been a return of popularity for the older established breeds, especially Brahmas, Dorkings, Game, and the old English Game. Certainly the day of breeds of such proved value is far from over.

Chapter 2
Selection of Breeds

When the beginner catches the "hen-fever"–probably at some show where the enthusiasm of fanciers is in itself contagious–and decides to keep fowls, his first question, put by himself to himself, is, or should be, What is my object in keeping fowls? That he may get a healthy amusement is certain; that he may find his hobby profitable is very probable, but the latter will much depend on the prudence with which he suits his selection of a breed to his accommodation and requirements. Some breeds are particularly good layers; others particularly good for table; some require large dry runs to do well; others will stand confinement and, in some degree, damp.

If he has good dry runs, and wishes to breed poultry for table, he will probably select some such breed as Dorkings, Houdans, Game, Indian Game, etc. If for eggs alone, Minorcas, Wyandottes, Orpingtons, Hamburghs, Polish, Spanish, Leghorns. If for both purposes, few breeds are better than Brahmas, or Langshans, and Brahmas crossed with one or other of the above breeds. For table purposes no doubt first crosses are best; same splendid results are so obtained. If a smaller bird of high quality is preferred, few crosses can surpass Game, or old English Game cocks crossed with

Dorkings, or Brahmas. If a larger bird be required, the cross of a Dorking cock with dark Brahma hens will produce a magnificent table fowl, and of fine quality. Langshans are also fine table birds, and would, no doubt, answer well for, crossing with Game or Dorkings.

If his object is to exhibit, and breed his own winners, he will probably consider only one point: namely, which breed he most admires of all those he sees at the shows, and will purchase stock of that variety to begin upon. Let him not be deluded by the foolish old saying as to the advisability of buying "unrelated birds." For mere farmyard purposes this may answer well, as greater vigour is probably obtained in this way. But for breeding for type for show purposes, and to establish a yard which may be trusted to breed true, such a plan is fatal. The prode of unrelated birds may be relied on with fair certainty to "splash" in various directions; that is, the young ones, instead of resembling the parents in their good qualities, will often be not only most unlike their parents but also most unlike among themselves. Nor in many breeds is it of any use to simply buy the best show cock and the best show hens available. For instance, in the dark Brahma, where, in the cock, pure white upper colour and black breasts are desired, these points are rarely, if ever, to be found in the fathers or brothers of the beautifully marked silver hens and pullets which it is the aim of the breeder to produce. In this and in many other breeds it is necessary to have cock-breeding pens to produce the cocks and cockerels for show, and pullet-breeding pens to produce the hens and pullets for the same purpose. The cock-breeding hens and pullets thus being only available for breeding, and so also the pullet-breeding cockerels and cocks.

More will be said on this subject as each breed is described and discussed later on. In any case, let him take care that he begins with fairly young birds–that is, not more than two years old. Two-years old hens and a cockerel of the year is, perhaps, best; next best, a two-year-old cock and pullets.

Here it may be well to define what is meant by the term cockerel and pullet, as distinct from cock and hen.

For show purposes, young birds are properly called cockerels and pullets during the year in which they are hatched. But from a breeding point of view, for a purchaser, they are often defined thus:

a cockerel is a young cock which has never moulted through completely, after having attained his full chicken feather, and is not in his full adult plumage.

The same definition may be used of a pullet as distinct from a hen. But in hot seasons it sometimes happens that early chickens moult right through, even when seven or eight months old; these must be regarded as still cockerels and pullets; therefore it may be well to add to the above definition, "a cockerel is a 'young cock,' and is for show purposes called 'a cockerel' during that year only in which he was hatched." For *breeding* purposes, however, a young cock counts as a cockerel until he goes into his adult moult in the summer or autumn of the year following that in which he was hatched. Hence it is obvious that a bird hatched in November of one year cannot be shown in the cockerel classes of the following year without fraud, though he may still be less than one year old, unless the schedule clearly specifies that the class is for "cockerels hatched during the previous year," which is done in some spring shows, and is quite intelligible and reasonable.

Chapter 3
Importation of Eggs

It is often said that it is a disgrace to the farmers, cottagers, and poultry keepers of England that several millions of pounds should be paid away to foreigners annually for eggs, which money might be easily kept in the country if English men and women would only take the trouble to pick up the money which practically lies at their feet. There is no doubt much truth in this view, though like many other truths, it is not the whole truth. The missing item is the difference of climate; the far dryer and, as a rule, warmer climates of Italy and France enable poultry keepers and others to get a good supply of eggs in autumn, winter, and early spring far more easily than Englishmen can do it. So far the advantage is with the foreigner, and cannot be gained from him by the Englishman. Another advantage the foreigner has for the most part now, is the system by which higglers regularly go round and collect the eggs from the cottagers and small poultry-farmers; thus ensuring a *large* and *regular* supply of fresh eggs getting into the egg-merchant's hands all the year round. This advantage should not be left with the foreign egg producer. There is no reason why systems of egg-collecting should not be made to pay well in England, and the system of co-operative egg-farming, which this means, would no doubt be as successful as Canon Bagot's co-operative dairy-farming experiments in Ireland.

On the other hand, there are two advantages which the Englishman has over the foreigner. He is nearer his market; and now that the railway companies are so greatly reducing the cost of carriage of eggs, he should be able to put them on the English market at a less cost of carriage than the foreigner. Moreover, he can bring his eggs to market fresher than those can be which come from abroad, especially those which have to travel many hundreds of miles, from Italy.

The cottager, who lives in the country, by the road-side, has a grand opportunity. His fowls have a good, free run, find a large part of their living, and are sure to keep in health. One thing ought to be done to make up for the damp climate and the long winter nights. The cottager or his wife should go to the hen-house with a lantern the last thing before bed-time, and give the fowls a warm meal by lantern light. They will soon look forward to it, and eat it readily, and it will keep them warm and well nourished till the morning meal. Here lies one of the great secrets of getting a good supply of winter eggs. Of course, for egg supply, only early hatched pullets and one-year-old hens which have moulted early should be kept. Each spring, when eggs get cheap, the hens should be sold off–for sitting purposes there is often a great demand–only the pullets (and perhaps one or two exceptional layers among the hens) being kept on for another year's work, the hens parted with being replaced by early pullets of the current year.

Chapter 4

Accommodation: Houses

The would-be chicken-keeper may have every opportunity and advantage for carrying on his hobby–unlimited dry grass runs shrubberies and trees for shade, a running brook for fresh cool water, strong and handsome buildings for his fowls, plenty of persons in his pay to look after the birds, keep them clean, collect the eggs, etc., or he may have few or none of these advantages. Though living in the country he may have but a small yard or a garden, of which he has to take a strip for his fowls; he may have to put up a wattled hurdle for shade; to house them in an old stable, or in a cheaply run up house of old boards; in an old shed, an old railway carriage, or even an old omnibus; and yet do well with his birds, perhaps beating his richer neighbour, who does not, or cannot, give the same amount of personal attention to his hobby as the poorer man, who is always on the spot and makes his poultry his chief personal amusement, to the great gain of his pocket and to the possible loss of the public-house keeper. He may even live in the thick of a great town, his country view limited to a small back yard, and he may be inclined to say to himself that under such circumstances he has little or no chance of success. But it is not so. Even under such circumstances, hundreds–nay, thousands–of persons have kept poultry most successfully; aye, and exhibition poultry, with which they have made a very valuable addition to the family income. In such a yard I should advise as follows: Build a lean-to house in one corner, so getting two sides of it contributed by the garden wall. Let it, if possible, be a corner facing south, or south-east or even south-west–Not north-east, at any rate. Make the side and end of the house of boards; good weather boarding is the best, to avoid any draughts. Let the roof have a good fall, and be built of stout weather boarding. I do not advise felt, as I find that, unless frequently tarred, the small birds, and large ones too, will pull it off for their own nest building. Let the door be in the corner of the house farthest from the wall, and leading into the wired-in run I am about to describe. In the door let there be put a slide, which should not go to the very bottom of the door, but some three inches from that–this enables one to have the door strongly made at the bottom, a very important point when that door gets old–and let the door open outwards and rest on a strong sill, some two or three inches deep. Then let the floor of the house be dealt with. If possible, a concrete floor at the bottom, to keep away rats, etc. If not, some chalk mixed with dry soil will make a firm and good bottom, which should be rammed hard, and set as firmly as possible. This floor should reach about up to the bottom of the sill, or a shade above. Then some three inches of dry ashes should be sifted on the floor, and these should be kept regularly raked, and fresh ashes added as required. On the most convenient side of the house should be the nest-boxes; and in most cases I should advise that they project out of the house, with a flap to cover them, which requires a padlock, and a slide to push up at night and withdraw in the morning, to prevent the birds sleeping in the nest-boxes, which will soon become foul and full of insects if so used. The slide should draw in and out horizontally, not perpendicularly. In front of the house, running beside the wall, should be as large a wired-in run as can be spared. If possible, it should be a fair height, say six feet; to enable a man to go about in it comfortably. If this is not possible, it should be made as nearly six feet as can be managed. Upright posts driven into the ground, and connected with the wall by others taken across either horizontally or sloping like the roof of the house, answer well. Perhaps strong iron rods curved to go right over the top of the run are better, but of course more expensive. Over this, wire netting must be carefully fixed, and it is good economy to have this fairly strong. The mesh need not be very small, 1½ in. to 2 in. answers well for ordinary purposes, though young chickens cannot be kept in by such large mesh. For a chicken run I should prefer 1 inch mesh. There should be a door frame, covered with wire, to match the rest, to give admission to, the run; and this wire door should be hung close to the shed, so that both doors are close together. The surface of the wired run should also be covered with either good gravel or sifted ashes. In any case, a handful of fresh gravel, thrown in every day, will be a great help to the birds, as it greatly assists their digestion.

If the birds chosen to keep are either Brahmas or Cochins, I should not advise perches, as these breeds, if the sleeping-place is dry, do quite as well, or better, without perches. But if the ground is at all damp, they also should have low perches, to be about fifteen or eighteen inches from the ground. In any case the perch should be of good size and rounded and, if a pole, should have the bark removed, or it will be a great harbour for insects. An occasional brushing with paraffin will be very useful for this perch; and when time can be spared to lime-wash the house well, with a little Jeyes' Fluid and paraffin added to the lime-wash, excellent results will follow.

For the lighter breeds perches are necessary, and if care be taken that they are arranged like steps, no one being above another, so that the droppings of birds on a higher perch shall not fall on those below, the more perches the better, as it gives the birds a means of exercising themselves. All ventilation should be at the top, and should be so arranged, by a slide or otherwise, that a great deal of fresh air can be allowed in summer, and a more moderate amount in winter. Even in winter want of fresh air causes the houses to become close and stuffy during the night, and then when the morning air comes to be breathed, colds and chills result, and are put down to the cold outside, whereas the real cause is the heat and foul air inside. "Humans" may take a useful hint from this. One or two small panes of glass should be let into the house-front on the sunny side. This is of advantage in three ways.

Firstly, light is of great importance for health, a fact often lost sight of by people who apparently try and keep health out of their homes by closely drawn curtains and blinds.

Secondly, it enables the birds to see their way about in the house at daybreak, and it is possible to leave them a little food–corn, not soft food–overnight, so that they will have an early meal before their owner need let them out, though if their owner is an early riser, and can feed them between five and six, I should prefer a hot meal of soft food, to corn left overnight,

Thirdly, for convenience when lime-washing, repairing, etc., when it may be necessary to have light inside, but the door shut. If the roof of the whole run can be boarded over like the house itself it will be a great gain, as the run will be kept dry and will be much more easily kept clean. Apiece of guttering to carry off the rain from the roof, which should be conducted to a drain or soft-water barrel, will help to keep the place tidy and dry. To the man or woman who starts fowls in such a back yard as I have described, I would add the warning, don't try and keep too many; and if you are going in for egg production you may safely have from six to twelve hens with one cockerel. Let your hens or pullets be early hatched. Buy forward birds about August, such as look likely to lay early, and prefer birds which have a good deal of comb for their breed. Such are likely to give you eggs in the winter, when eggs are scarce and dear, and here is your best chance of a good profit. Feed well, vary the food, but don't let your birds get too fat. A stump of a tree placed in the run for them to fly up and down is a good help to exercise, and therefore to health.

Where movable nest-boxes are used–and next to boxes projecting outside the house they are much to be preferred–they should be taken out of the house each night and placed in the dusting sheds, that the birds may not sleep in them and foul them.

Various forms of movable houses are used, and trade advertisements of such are very common. Some of the larger ones are on wheels, and can be easily moved by a man, or in case of very large ones, by a pony or horse. Smaller ones are made with handles and poles, like the old-fashioned sedan chairs of bygone days; others with a wheel or wheels in front and handles behind, resembling a sort of wheelbarrow. All these are useful for their purposes, and where there is plenty of room available, it is a great gain to be able to move the houses about, and thus obtain fresh sweet grass, and a fresh supply of insect food.

Chapter 5

Marking Fowls and Eggs

It is often suggested that marking-rings, such as are used for pigeons, should be regularly required for exhibition poultry; but, with most breeds, there seems no likelihood of this being successful. The growth of a pigeon's leg, between marking time and its adult growth, is not so great as in the case of a fowl; and it is very possible to have a ring for a pigeon which will just slip on at marking time, will keep on, and yet not be too tight when the bird is fully grown. In some, and probably most, breeds of fowls, this is impossible. Any marking ring small enough to put on a young Brahma, would be much too, tight on a big cockerel. Any ring large enough to be wearable on a big cockerel at ten months, would be useless for a young chicken, and, by the use of soap and other devices, could be got on a partially grown Brahma, long after he ought, from his age, to be marked.

I mark every chicken as soon as it is hatched by twisting two strands of *single* Berlin wool and tying it round one leg of the chicken, leaving, after one turn of the wools, a loop, such as perhaps a pea could pass through, before further tying. I then tie fast several times, cut off the ends with the scissors, and if properly done, that chicken is marked for many weeks. When necessary, the operation is repeated with a twist of *double* Berlin wool of the same colour, leaving again a loop for growth, as before. This generally lasts till the birds are six or seven months old, when the process is repeated on a still larger scale, and lasts for a year. Each chicken takes only about a minute to mark, and as many colours can be used, and the marking on the right or left leg may be used to denote different pens, it is easy to have a score of different marks for one's birds, which can be removed when required for show, and leave no permanent mark on the bird. It is desirable to remove coloured marks before washing a bird for, show, or the colour may run, and leave an ugly stain on the bird's leg.

I always mark my eggs on the *pointed end*. I may add the mark at the other end but the pointed end mark is the essential one. This is not the usual practice, I believe, but I found that as chickens nearly always chip at the round end when hatching, the mark, if at the round end alone, is destroyed, and where mixed settings are used, it is difficult, if not impossible, at times to tell how many of each have hatched, especially where an egg or two may be broken.

Other plans of marking chickens are by punching holes in the web of the foot, or in the web of the wing; or an elastic band may be placed round the wing, close to the body, and will be quite unnoticed as the feathers grow over it. This latter is a good plan in many ways; but on the whole I prefer my wool arrangement for quite young chickens. When the youngsters grow as large as a half-grown partridge I now add the india-rubber band on the wing, as an additional clue to identity, in case the wool on the leg should wear or drop off. The india-rubber band should be placed on the corresponding wing to the leg, which is marked with wool, right or left as the case may be; and a little wool of the same colour as the wool on the leg, tied loosely two or three times round part of the band, will for years be available to identify the bird. The band first put on will last till the bird is nearly full grown, when a larger band can be substituted, which will probably last the bird's lifetime; while the small band which has been taken off will do again for other birds. Every egg should, when taken from the nest, be marked with the date of laying; and if more than one pen of fowls is kept, with the name of the pen. If more than one breed is kept, some such plan as the following may be adopted.

When I kept both dark and light Brahmas, I marked all the dark pens by letters of the alphabet, A, B, C, D, etc., while the light pens were marked numerically, 1, 2, 3, 4, etc. This plan saved confusion, and could of course be developed if necessary.

In setting mixed settings of darks, I found that by setting cock-breeders and pullet-breeders together, only from one pen of each in each setting, of course, I could make up a setting more quickly; when the chickens fledged it was easy to tell which were from which strain; but those I was not sure of at the time of hatching I marked as *both*.

Thus, if the cock-breeders should be marked red on right leg, and the pullet-breeders yellow on same leg, I twisted red and yellow wool together. Thus I knew they were from one of two pens, and as they fledged could tell for certain from which; the doubtful mark is then removed, and the one colour alone placed on the leg.

Chapter 6

Management and Feeding

In fine weather, and throughout the summer, it is very important that the fowls should be let out early, and that when let out they should be fed soon. This is especially important on cold mornings, and in cold weather, as the system soon gets chilled in the morning air after some hours without a meal, in the comparative warmth of the fowl house.

If any of the well-known poultry meals sold by various firms be used, it is best that they should be used diluted with a fair proportion, at least half, of ordinary meal. In that case the process should be as follows. Take the required quantity of the patent meal and half fill a bucket, pour boiling water on it, so as to fill the bucket, and place upon the bucket a board or cover, so as to retain the steam. In a short time the meal will have swelled considerably, when the hot mass should be turned out into the mixing tub and well mixed with the meal to be added, till the whole is of equal consistency, and in a moist *crumbly* condition. If available, good soup in which dogs meat, etc., has been cooked, is excellent for mixing the soft food, and the meat itself, broken fine and mixed in well, is of great value. Both these are more suited for winter work than for summer, when they are rather too stimulating: If the poultry keeper can pick up, cheaply, parts of a horse which has had to be killed, and use it as above advised for his fowls, he will find it both good for the fowls and economical. A bone with a little of the meat, whether cooked or raw, left on it, and put in the run for the fowls to peck at is also most beneficial.

The maggots from a tallow chandlers are also valuable food for poultry if used in moderation. Barley-meal-and ground oats make a splendid meal for fowls, especially if a little middlings be added. Equal parts of each, mixed with boiling water, and mixed rather crumbly, so that it breaks when thrown on the ground and is not sticky, is as good as anything in the meal line.

If well-cooked vegetables, scraps of meat chopped fine, bone-dust, house-scraps of any sorts can be mixed with it, it will be improved. Fowls, like other creatures, are the better for some variety in their food, which the house-scraps tend to provide.

In summer, with a good run, they need have nothing more than a few scraps at mid-day until the last meal at night, which should always be of whole corn. As a staple food I prefer wheat, and so do the fowls. Good barley and oats are a useful change. Maize should be seldom used. It contains too much heating and too little flesh-forming material, and tends to cause unhealthy fat and disease of the liver. Buckwheat in winter is a nice change, and is good as an occasional food for young chickens. In winter the mid-day meal for adult fowls must be increased into a fairly substantial one, as more food is required to keep up the warmth of the bodies, and in winter there is little or no insect food which the birds can pick up for themselves.

A little chopped meat daily, during the winter months, no doubt adds to the number of eggs, and also to the proportion of fertile eggs among those laid, as it takes the place of insects, which are not easy to provide for birds in small town enclosures.

Avoid foul water for the birds, and especially sun-warmed water in the heat of summer. The water vessel should in summer always stand in the shade, but in winter in the sun if there is any.

Some sort of grit is absolutely necessary to the digestion of fowls. A handful of gravel thrown among them daily is excellent; pounded oyster shells are also very good, and in addition provide plenty of lime for the formation of shells. The oyster shells should be baked in the oven, when they will be found much easier to pound up.

Even broken crockery, pounded small, about the size of peas or smaller, has been found very useful by people who could not easily get grit. Road sand answers capitally, and makes a very good floor for the house or run.

Vegetables of some kind daily are most necessary; cooked cabbage, chopped fine and mixed with the food, fresh lettuce leaves, in winter a piece of mangold, a Swede turnip for the birds to peck at, a sod cut from the roadside. All these and others are available to keep the birds in health. Of course, in a grass run they are not necessary, though even there change is good.

Feeding for colour, that is, giving chemical or other foods to produce increased depth or brilliancy of colour, has long been practised in the cage bird department of the exhibition world–whether advantageously or the contrary may be doubted. Now, with the late "buff craze," the poultry world is face to face with the same question, and opinion is strongly divided as to the rightfulness or wrongfulness of "colour-feeding." There are also some who hold that colour-feeding, so far as it is known at present, is a failure, and that it does not produce any improved colour to the advantage of the exhibitor of the colour-fed specimen. Again, at present, no chemical test is known which can detect the colour-fed bird from the naturally fed; and those who wish colour-feeding prohibited by the Poultry Club, are met with the reply that while detection is impossible prohibition is ridiculous.

Again, the question is raised, "Where are you going to draw the line? Maize increases the richness of yellow or buff plumage, and gives a more yellow leg. Iron adds brilliancy, so does boiled linseed mixed with the food; are these also to be prohibited?" It is obvious that iron is a tonic, and linseed a seed, and therefore a natural food for birds, and so far distinct from mere colouring chemicals. But the whole subject bristles with difficulty, and is at present hardly ripe for decision, though one of the largest specialist clubs has carried a resolution against colour-feeding in any form, a resolution which does not include such things as iron and linseed in its prohibition; as these, though they may give brilliancy, do not alter the tone of any colour.

Chapter 7
Breeding Pens

For breeding exhibition poultry, careful mating, namely, careful selection of the cock and his mates, to correct the faults on the one side by the excellences of the other, or at any rate not to "breed in" to a bad fault is necessary. And to get strong chickens it is desirable that an excessive number of hens should not be run with the cock. If eggs alone be the object it does not matter, as the question whether the egg would have hatched into a strong or a weakly chick matters little to either the vender or eater of the egg. But for rearing good stock, this question is of the greatest importance.

And in the first place it may be pointed out that in the cold weather birds are not so fertile as in the warmer months, and that as a rule cockerels are far more fertile than cocks. Also the heavy breeds, such as Brahmas, Cochins, Dorkings, Langshans, etc., should not have so many hens in one pen as the lighter breeds, such as Hamburghs, Minorcas, Wyandottes, Leghorns. Hence it follows that in the early part of the season it is most unwise to overcrowd the breeding-pens.

For heavy breeds it might be suggested:

1. *For a cockerel*: till end of February, 3 hens; then increase to 4 or 5.
2. *For a once moulted cock*: early, 2 hens; later, 3 or 4.
3. *For an old cock*: early, 1 hen only; later, 2 to 3.

For the lighter breeds:

1. *Cockerel*: early, 4 or 5 hens; later, 6 to 9.
2. *Cock*: once moulted, early, 3 or 4 hens; later, 5 to 7.
3. *Old cock*: early, 2 hens; later, 4 or 5.

These numbers are only given as suggestions and approximate. The individual birds vary so much that observation will at times give reason to change the numbers. If the hens are observed to be losing their feathers on back or sides, it is generally wise to add another hen or two. Fresh birds added should always be placed in the pen after dark. There is less likelihood of a severe fight when birds wake up together in the morning. As to varying conditions of fertility, one of my experiences with a great winner may be of interest and worth noting.

I put the best hens I had with one of the best cockerels I ever bred. He never fertilised an egg as a cockerel! He won a Cup at the Crystal Palace as a cock. I again gave him some of my best hens, but got very few chickens Again he won a Palace Cup. In despair I gave him again a few hens, but none of my best; he fertilised almost every egg in this, his third year, and one of his sons bred a Palace Cup cock for me. The old cock again won a Palace Cup for the third time, and I sold him for exhibition purposes.

Chapter 8
Setting and Hatching

For setting hens I prefer detached boxes, which can be easily moved, lime-washed, and generally kept clean, to fixed rows of nests, in the chinks and joints of which insects are likely to congregate. I use empty boxes, such as orange boxes, divided into three parts (or if small into two parts) by wooden partitions. The top should be removable, and should allow for plenty of ventilation. The fronts should take on and off, so that the sitting hen can be shut in, and daily let off for a quarter of an hour to feed. If let off morning and evening, it will be all the better.

Each division should be large enough for a large hen, and deep enough to allow of a good deal of ashes and straw for the nest, and a good air space between the hen's back and the top of the box. At the bottom put a good lot of *finely* sifted moist sand or cinder ashes. If not finely sifted, as the incubation goes on, pieces of hard material are likely to work through to the eggs, and if the hen then treads on an egg which touches one of these bits of stone or coal, the chances are that the egg is broken.

On the top of the ashes; which should be scooped into the form of a nest, add plenty of good straw, and on that, especially in cold weather; a little hay. Then a few china eggs should be put in the nest, and *at night* the broody hen should be placed on the nest and shut in. In a few hours, from twelve to twenty-four, it will be seen whether the hen intends to sit quietly. The real eggs should by no means be put under her till she is steady and reconciled to her nest. Then, at night again, the china eggs should be quietly removed, and the real ones substituted. It is a mistake to put too many valuable eggs under a hen, especially in severe weather, when an egg, if pushed out from under the hen, soon gets chilled. In winter nine to eleven eggs are enough, later on thirteen may not be too many, and in very hot weather a large hen may be trusted with even fifteen eggs. It is a good plan to sprinkle a little pyrethrum powder under the hen's wings and about her back and fluff before setting her, and again when she hatches out before putting her in a coop with her chicks. The eggs should be carefully tested by candle light when they have been sat upon for a few days. Some can test earlier, but for most people the seventh or eighth evening will be early enough. All clear eggs should be removed, and can be used for puddings, etc., for which purposes they are excellent. The clear eggs can be seen through when held before a candle, much the same as a freshly laid egg. The fertile eggs will be opaque, or clouded, and should be carefully replaced without shaking. The lessened number of eggs is more handy for the hen on the nest, the eggs get more certain warmth, and are less likely to be pushed out into the cold.

Some prefer to use artificial incubation, and it no doubt answers well for those who will give ordinary care to it.

The incubator should be placed on a *firm* stand or table, in a well-ventilated place, free from draughts. The thermometer should be tested for a few hours before the eggs are placed in it, and if steady at from 102 to 104 degrees the eggs may safely be put in. For incubators it is very important that the eggs should be fairly fresh. If possible no eggs should have been laid more than a week or ten days before they are placed in an incubator. They should be turned daily, though perhaps not necessarily turned right round. They should be moved to vary the positions in the drawer. Thus those in the centre should be placed at the outside, and *vice versa*. The drawer should be left out for some little time morning and evening to air the eggs; the neglect of this is most injurious, and there is no doubt that plenty of fresh air is of great importance if strong chickens are to be obtained. In winter, and on cold days especially, when spring *east* winds are blowing, care must be taken not to leave the eggs out too long, or the germs may get chilled. In warm weather a quarter of an hour is not too long, especially when the eggs are due to hatch in a few days.

Since writing the above I have been much struck by a letter in one of the Poultry papers from one who has had much experience of incubators. He emphasizes most strongly the importance of avoiding sudden shocks, jars, noises, etc. Prefers the heat of 103 up to the sixteenth day, afterwards reducing it to 101 degrees. And airing extended when near hatching even up to one and a half hours at a time. Of course the avoidance of draughts is absolutely necessary when the eggs are out for an airing. To avoid this, he covers the eggs with porous sacking or some such material, and allows his eggs to get almost cold before he returns them to the incubator. He adds one more instance to the many recorded of the wonderful vitality of the egg–or rather its occupant–especially towards the end of the period of incubation. He says: "Last week, with a cold east wind blowing, I forgot to return one of my drawers, containing 105 eggs, which was left out for *quite four hours!* I thought this time I had overdone it, so opened an egg and found life still there. The hatch was the best I have had this season, every egg turning out its inmate on the twenty-first day."

This case is most encouraging, and if this experience is generally verified by others, goes to prove that the difficulty with incubators has been caused principally by insufficient airing or draughts on the egg. He also thinks that a temperature of over 105 causes cripples, as the chicks turn round to get fresh air, and I suppose are unable to get back into proper positions. This, however, seems a case as yet of "not proven."

There is no doubt that the germs are more delicate from the sixth day till about the fifteenth than either before or afterwards. In fact, many cases have occurred in which eggs nearly due to hatch have got apparently quite cold, after even hours of chilling, yet have recovered their warmth and eventually hatched out strong chickens, though a little late. I have known a chicken hatch out six days after it was due; set under a hen, and; so far as could be discovered, without any reasonable explanation; but as a rule eggs overdue two days contain dead chicks.

It is very important that a sufficient amount of moisture should be given, whether the hatching be by incubator or hen, especially a day or two before the chicks are due. In the case of the incubator the tray underneath gives the opportunity to add a little lukewarm water. When a hen is used, it is desirable in dry weather to pour a little lukewarm water round and upon the nest. The steam will rise when the hen goes on again, and will keep the skin inside the shell fairly moist. If this skin gets really dry the chicks cannot penetrate it, or if they do succeed so far, it will adhere to them, prevent them from turning round inside the egg, and thus prevent them from chipping enough of the shell to get out. The chicks usually emerge from the large end, where the air bubble is, but some are turned round, yet succeed in hatching from the narrow end.

The chick absorbs the yolk, which has been kept in reserve for it all through the process of incubation, just before it hatches. Some chip and get clear before they have done so; these almost always die, but if not quite out a few succeed in absorbing it at the last moment.

Eggs a day overdue should be tested by floating them in a vessel of lukewarm water–not hot, but just so warm as to feel comfortable to the hand. If allowed to become quite still on the surface of the water, the live chickens will cause the eggs to give odd little twitches or jerks, and these are likely to hatch. Those in which no movement can be seen are usually dead, or so weak that they would not hatch.

It used to be said that if it became necessary to help a chicken out, and in the process it bled, the loss of blood was fatal to it. This is not so, and I have known hundreds of chickens which bled considerably while being helped out of the shell prove good strong birds.

It is best to hold the greater part of the egg in warm water while helping the chick, taking care to keep the head above water. If the head, neck, and shoulders can be got fairly clear, the chick will generally struggle and help the operator by its own efforts.

Chapter 9

Coops, Artificial Mothers, Food and Rearing

When the chicks are hatched, they must either be placed in coops with the hens which have hatched them, or have hatched others, or under the contrivances which are called artificial mothers. This should not be done till they are twenty-four hours old, as it is necessary that they should get thoroughly dry, either in the nest or in the drying-box of the incubator before being exposed to the fresh air and risks of chill. During this period they need no food, as they have the yolk–the most nourishing and digestible food possible–to sustain them for the first twenty-four hours. The queer old practice of giving each chick a peppercorn seems both unreasonable and likely to be injurious. I never heard of anyone who proposed to give newly born babies or puppies any hot pepper as a comforting start in life, and I cannot believe it is good for chicks.

When first put out they should have a meal of some wholesome and. digestible food. For this purpose I would strongly recommend dry bread-crumbs, mixed with a little finely chopped hard-boiled egg. Care must be taken to only give chickens as much as they will eat up clean. Sour food left lying about means disease and death. A little finely chopped lean meat every day is most beneficial, and if they cannot run to grass for themselves, a fresh sod by the coop is excellent.

I prefer to give young chickens alternately soft and hard food, but always hard food for the last meal, which should be given as late as there is light for them to see to eat. At first, five meals a day are not too many, but gradually they should be reduced to four, then to three, which is the best number when they get nearly grown up. For hard food I prefer, for the first week, "grits" and split wheat, but they must not be allowed to over-eat themselves with wheat, which they will sometimes do. After a week, I drop the grits and use split wheat only, till they are old enough to have whole wheat. I prefer wheat to all other kinds of grain, and so, I find, do the chickens. They should be let out as soon after daylight as possible, and fed with soft food directly they are out, or at any rate within a quarter of an hour.

Take care that the water is quite fresh, and the drinking vessels kept quite clean. Above all, beware of sun-warmed water in the hot weather. As chickens, which have been shut in all night are usually thirsty when first let out, it is important that the attendant, or whoever lets them out, should take round a can of fresh water, and, having thrown away the old stale water in the vessels, fill them afresh. For young chickens I like flower-pot saucers glazed inside. These keep very clean, and are not expensive.

It is essential that a coop should be watertight above and below; should have a front for night time to protect the chickens from rats, weasels, cats, etc., and plenty of ventilation without draught.

It must also have a strong, movable tray, fitting inside to form a floor to the coop, which should be covered with fine sand, ashes, or gravel. This can be daily raked over, to keep the coop clean and keep down insects. Before putting a hen in a coop with chickens, sprinkle her well with pyrethrum powder, especially under the wings and all the lower fluffy parts. This may be repeated when necessary, and the chicks may also be treated in similar fashion. They will not grow if tormented and partly devoured by insects; and lice are even more trouble than fleas, and more often unnoticed. Examine carefully the young chicks at intervals, and especially under wings and at back of head. A good heap of dry soil, gravel, or ashes for them to run to and dust in will be of great service.

If artificial mothers are used, care must be taken that the chicks do not get entangled in the flannel, or whatever is used as a substitute for feathers. In the early days a good many chicks got hung up in this way, but most mothers are now arranged to avoid this. It is as well to say, *keep the heat in the mother too low rather than too high*. Excessive heat means chills and deaths. Plenty of nice, small gravel thrown down for the chickens in the run is most important, giving them occupation and helping them to digest their food.

Again, do not let access to green food of some sort be forgotten, and fresh, clean water. See that the floor of the mother is cleaned every day, and thoroughly. Look out for insects, and if you find any, act at once. Paint woodwork with paraffin, and sprinkle pyrethrum powder on chicks, floor and wherever necessary. If the chicks seem to have a difficulty in growing their feathers, a little flowers of sulphur mixed with the food, say twice a week, is very beneficial. It is also good for adult birds when moulting, to help the new feathers. For leg-weakness, iron in the water and bone-meal in the food are excellent.

Chapter 10

Preparing for Show

When fowls have a good grass run, little preparation for showing is necessary beyond occasional penning to get the birds used to the show pen. Most birds, when put into a pen for the first time, are wild, and refuse to show themselves. Under such circumstances they would have no chance against birds which, though not really so good, were quiet, and stood up, showing themselves off. If birds are penned for a day or two, fresh green food and clean grit of some sort should be put into the pen for them; this prevents many of the bad effects of penning. Of course birds have lived in pens for months, but great care and skill have to be applied, and, under any circumstances, it must be most injurious to the bird.

In the case of Game and Game Bantams a somewhat more elaborate system of penning is pursued. The birds are put in pens for longer periods and regularly trained, by holding food up high for them to strain after, to show the "reach" which is so admired in the breed. But for birds where this "reachiness" is not desired, a much shorter system of penning will do well.

After occasional penning the birds need only be put in the pens about four days before the show, when they should be carefully washed, and as carefully dried. The main point in washing a bird is:

1. To get it thoroughly wet, and the soap well into it, to the very skin.
2. The far more difficult matter of getting the soap *well out*.

Before washing, it is well to give the bird a teaspoonful of whisky and water–about one-third to one-half whisky. Then wash well in lukewarm water–if curd soap be cut up and boiled in a little water it answers better than if used hard. When the bird is thoroughly clean rinse *well*, first in lukewarm water, slightly blued; then again in cold water, also slightly blued.

Great care must always be taken to stir round the blued water immediately before dipping the bird, as the blue has a lay of settling at the bottom of the tub, and I have seen many birds, in other respects well washed, which were quite spoiled for show, as their feet and legs were a bright blue, they having, no doubt, stood in the sediment at the bottom of the tub and become really dyed. It is by no means easy to get this blue off again, so the caution is very necessary. Should, unfortunately, any bird have a fit from shock or excitement when being washed, I have found the best plan is to let cold water from a tap or pump trickle on to the head, though the deaths of two celebrities under the washing process were very disheartening. Since, however, we have given whisky before washing, as well as after, we have had no collapses.

When the bird is thoroughly rinsed, the water should be squeezed out or got out with a sponge as far as may be at the time, and then the bird should be put into a hamper (lined), which must stand in front of a hot fire. Care must be taken that the hamper cannot catch fire. When the bird is about half dry it is a good plan to stand it on a plank on top of the basket (taking care that the bird is not scorched), and by holding up the feathers so that the heat of the fire can penetrate, get the fluff and underparts well dried. This is the time to see that the hackle is drying straight, and a comb may be made useful to hold up feathers so that the warmth may get to the skin.

For a week or ten days before showing, a little linseed, boiled and mixed with the soft food, will add to the gloss of the feathers. When the bird is thoroughly dry it should be placed in a clean show pen, the floor of which has been covered with clean straw chaff. This must be kept thoroughly clean until the bird is sent off to the show, and frequent inspection will accustom it to visitors, and prevent its being shy when the judge comes round. A little green food should be given it every day, some small stones or grit, and a few bits of charcoal placed in the corner of the pen will be found useful. Before sending off, gently rub the comb, deaf ears, and wattles with the finger dipped in a mixture of equal parts of vinegar and sweet oil. This brings up the colour and freshens the appearance. Be careful not to daub it on, as a little answers every purpose.

Chapter 11

Fattening, Killing and Plucking

To those who, having a good run for their poultry, are considering the advisability of putting some of them up to fatten for their own use, I should offer the advice proffered by Mr. Punch, on an almost equally momentous occasion, namely, "Don't."

I believe the flesh of a healthy fowl, which has lived a free, out-of-door life till the last moment, is both better in flavour and more wholesome than that of one which has been kept in confinement and fed, perhaps compulsorily, into an unhealthy condition of obesity. And if well fed, and killed at the right time, such naturally fed birds will be quite plump enough. Pheasants and partridges come fairly plump to table, even when left quite free to find their own living.

Occasionally it is advisable to artificially increase the flesh of birds to be killed for one's own table. In that case, if a few birds are confined in a clean, warm pen or run, and fed on as much soft food as they will eat–say three or four times a day, care being taken to feed very early in the morning and as late as may be at night–the birds will gain in weight, and will eat well. It often happens that for the first few days no great increase of weight is to be noticed; occasionally for a day or two a loss of weight occurs, owing probably to fretting, but after the first few days the gain in weight should be very noticeable.

But where special fattening for market is desirable, a few of the ordinary rules or principles may be given.

Firstly, to fatten successfully, the birds should be kept in the dark, or, at any rate, in a dim light, which will encourage them to rest, an essential to the laying on of flesh.

The fattening pen, whatever sort be used, must be kept scrupulously clean, and be well time-washed beforehand, and, if necessary, at intervals, with ordinary lime-wash, to which a little size and a little Jeyes' Fluid or petroleum has been added, the former for choice.

The birds should not be fed at all till they have been in the fattening pen some few hours, long enough to get hungry, and while they are being fattened must have only soft food, no hard corn being admissible. The floor of the pen should be made of strong laths or strips of wood, so that all manure may drop through and the pen thus remain clean. The pen should have a movable front, reaching nearly to the top, so that when it is fastened up the birds are almost in the dark. The fixed front behind this movable one should be barred; either strips of wood, or rods of wire, fastened vertically.

For fattening on a large scale, a machine I saw some years ago at the Jardin d'Acclimation, in Paris, which, no doubt, is still in use there, struck me as being very ingenious. Some hundreds of single pens were arranged, radiating from an upright iron pillar. The whole set were hung on the pillar, so that all could be turned round at once by a mere touch. The feeder had a seat on a mechanical contrivance, which enabled him to raise or lower himself a step at a time, using, I think, his foot for the purpose. He could thus start at the top tier of fowls, and by turning the pens by hand, feed all in that row; then lowering himself a foot, feed the second row, and so on till he reached the bottom row. The food was sufficiently liquid to pass easily though an india-rubber tube, just like a garden watering arrangement; the mouth-piece had a sort of gauge, so that by opening the mouth of each bird, and pressing his foot on a spring he could *squirt* the requisite quantity of food down the throat of each chicken. Curiously enough, the chickens seemed to look forward to and like it, though at first it must have seemed a most uncomfortable and terrifying experience.

For fattening foods, no doubt, the best are ground oats, buckwheat-meal, maize-meal, and whole wheat-meal. At any rate, the buckwheat-meal should form an item in the process. Some add suet and other fatty substances. These, no doubt, contribute to the fat of the birds, but not so much to the flesh, and in *fattening* fowls, as it is called, the true object is not to lay on *fat,* but to develop plenty of good, wholesome flesh. Take care that no stale food is left; this applies to the feeding of all animals or birds, but particularly to young chickens and fattened fowls, as soft food left about turns sour, is injurious when eaten, and gives the birds a distaste for food. No care is too great to take to scrape away or remove any soft food left over. The water, as I have before said, must be fresh and clean.

It is not wise to try and fatten too many fowls in one coop. The French plan I have before referred to keeps each bird separate, though they can see each other by sticking their heads out of their single pens. But for ordinary purposes I should prefer to have two or three cooped together, as birds are generally more quiet when they have companions to which they are used. It would not be wise to pen together birds which are strangers to each other; the first result would be probably a free fight, which would discount the effect of several days, fattening. If, however, it is necessary to put strange birds together, it is best, not only for fattening purposes but at all times, to make the change in the evening after dark. The birds hear each other, more or less, they wake up finding their neighbours by them, and probably, in the dim early light, in many cases get reconciled to the idea before they can see well enough, or are wide-a wake enough to fight.

Again, variety of food is desirable for every one, and for fowls as well as humans. To increase the appetite of the fattening fowl, it is especially necessary. The longest period that I should advise for fattening a fowl is three weeks. It will not lay on, so much flesh proportionately after that time, and the confinement and the unnatural system of feeding will begin to tell on the bird after three weeks.

If milk is plentiful, whether skimmed or new, it is a very valuable addition to the food, which should be mixed with hot milk, instead of hot water.

One of the most unpleasant parts of keeping poultry is the necessity for killing. No person of ordinary kindliness or humanity relishes the idea of hatching and rearing a large number of living creatures of great capacity for suffering, solely with the idea of bringing them to a painful death. As, however, animals must be reared and slain for the food of man, it is at any rate right to plead that the killing process shall be carried out with as great rapidity and as little suffering as may be possible. One shudders at the thought of the horrors of the slaughterhouse, where too often exists a sickening cruelty and a callous disregard for suffering, which is a terrible blot on our so-called civilization. Let us, at any rate, who keep poultry see that our fowls shall be killed with as little cruelty as may be.

The best way of killing a fowl, if one has a strong wrist, is to take the head of the fowl in the right hand, the neck in the left, and with a sudden wrench backwards of the right hand holding the head, to dislocate the neck. If this cannot be managed, I should advise the operator to tie the legs and wings of the fowl together, and with a sharp chopper chop the head clean off, chopping of course on a solid block. The tying is recommended because even after the head is off the muscular action is considerable, though no doubt without any sensation remaining in the bird. I have myself seen a headless fowl fly away for seven or eight yards; a gruesome sight which the tying renders impossible. The slow tortures which some other methods involve are a conclusive objection; a severe blow with a stout stick on the back of the head, if well managed, is a good plan, as if properly dealt it should cause instant insensibility. Plucking should be done while the bird is still warm, but not, as in some recent horrible cases, while the bird is still *alive!* The bird should be left till it has ceased to quiver, so that there can be no doubt of its death, and then the sooner it is plucked the better, and if necessary singed. Those who like the effect can then place it between two boards "to keep the breast down" as it is called. Some beat the breast bone down for the same purpose. As really it does not, alter the proportion of meat there is on the breast, I never have it done for home consumption. For market it may be well to concede so much to custom.

Some recommend that the bird should, when plucked, be at once plunged into boiling water, to improve its appearance. There is no objection to this, though it is only its appearance that will be improved.

If the poultry keeper also keeps pigs, he will probably give them the entrails from the fowls he kills. There is no doubt piggy likes them, and they are nourishing food which do no harm to the pig.

The feathers are a valuable item in poultry keeping, if carefully kept and properly treated. I have known a cook, who was allowed the feathers from the fowls she cooked as a perquisite, to be possessed of most excellent pillows, bolsters, and feather beds, which came in very handy when she was prepared to start on her matrimonial career. Her process was fairly simple, and was much as follows. Avoiding all dirty feathers, she took all the small feathers from back, breast and fluff, cut the hard ends off, and then at night, when her work for the day was done, she put them in the kitchen oven, where they remained all night, and were thoroughly baked by the morning. They were then collected into a bag, and kept dry, until she had enough for making the pillow, bolster, or feather bed, on which she was at that time intent. She might have also used the wing and tail feathers by stripping the feathery part at each side and rejecting the quills. If the feathers are collected for market they should be divided into classes–ducks' feathers apart from fowls' feathers, white apart from coloured, the strippings from the larger feathers apart from the smaller and more fluffy leathers. They should be kept in muslin or canvas bags which are exposed by their open texture to the action of the air. If hung up in the air on dry, warm days it will help to keep them sweet and dry.

Chapter 12
Cooking

Though this is not exactly a cookery book, it may not be out of place to give a few of the uses to which we put young and old fowls which are not required for breeding, or for sale as stock-birds.

As soon as the young chickens are sufficiently developed and fledged for it to be possible to make sure that they will not be wanted, they are killed, and made into what we call "pigeon pie." If treated exactly as young pigeons would be for the purpose, and a small piece of steak or bacon and pieces of hard-boiled egg be added, with seasoning, a most savoury and nutritious pie is the result.

Or they may be used in a similar way for a chicken pudding. A couple of sheep's kidneys, a few mushrooms and oysters may be added if it is desired to further improve the flavour. In the case of the pudding, the flesh, liver, and heart, and, for those that like it, the gizzard (properly cleaned first) should be used, but the bones should be omitted.

If half-grown birds are taken straight off the runs and prove not to have much meat on them, as does occasionally happen in the growing stage, they had better be boiled rather than roast. This is more often the case with young cockerels than with young pullets, which have generally flesh enough for roasting.

With regard to old birds, I have elsewhere referred to the two methods we prefer for rendering them more tender. The one, to half boil before roasting; or, if necessary, to boil them only. In both cases the word "boil" is a misnomer. They should be allowed to simmer only, and never allowed actually to boil; and this is one of the great secrets of rendering an old bird tender. I have known an old hen, which, if roasted, would have been uneatable, come quite juicy and tender to table, after a process of simmering for some three hours.

Old birds should be hung some days before they are eaten, though in summer the time is, of course, cut shorter. Or they may be wrapped in a napkin or cloth, and buried in the ground for twenty-four hours, when they will be found far more tender for the process. I never have an old bird roasted; the motto is, Boil, and boil long and slowly. If doubtful whether a bird will prove tender enough for roasting, it is an excellent plan to partly boil it first, and then roast. Most cooks are, no doubt, aware that to boil a fowl fast makes him leathery; a slow process, long continued, is what is wanted. Fowls should not be "drawn" if wanted to keep, but should be hung up by the legs head downward, and not touching each other. Ducks and geese should, of course, be hung up by the head.

Another favourite process with a very old bird is to convert it into *brawn*. Chicken brawn for breakfast is not a thing to be despised; and if well made and well seasoned, is a food for an epicure, even though the venerable fowl may have for some years braved the battle and the breeze.

The recipe for this I herewith give; that it can be varied I have no doubt; but here questions of taste come in. I eat it with mustard, vinegar, and pepper stirred together in my plate as a sort of sauce; some dislike vinegar, and eat it with salt, or salt and pepper only. It also goes well with a slice of cold ham or cold bacon.

Recipe for Chicken Brawn

Stew long and slowly till the bones slip out of the meat. Cut up the meat into moderate-sized pieces, say as large as two mouthfuls, add pepper and salt to taste, and fill one or more moulds with the meat, mixed with hard-boiled eggs cut into three or four pieces. While putting in the meat and eggs, add a little of the liquor the fowl has been boiled in; then add to the liquor which is left, half an ounce of gelatine, and when this is melted, pour over the meat in the mould as much of it as the mould will hold. Put the mould in a cool place to set, and when required for use, turn it out on to a dish.

Chapter 13

Enemies of the Poultry Yard

The worst enemies the poultry keeper has to deal with are rats, cats, weasels, stoats, foxes, and jackdaws. In some counties rooks are also found to take young chickens; but, so far as I have investigated, principally in dry localities where insect and grub-food is scarce. My own chickens are reared under a large rookery, but I have never lost a chick through rooks, though many have been taken by jackdaws.

Rats must be trapped, or with great care poisoned. Three-quarter inch wire netting sunk a few inches to a foot under the soil of the shed, and fastened up the sides of the house for a few inches is very effectual in keeping rats out. Cats which have once acquired the habit are very incorrigible, and I can only recommend murder where there is no means of keeping them out. Where the chickens are in a comparatively small yard, wire netting carried all round the walls or fences, and overhanging at each side, is most useful, and may be used in various ways to keep the cats out. In country runs, weasels, stoats, and foxes are often troublesome. Rats, weasels, and stoats one year accounted for some sixty of my best chickens. A sharp look-out with a gun was the only plan, though if small drainpipes be laid about, master weavel may often be found ensconced therein.

As for foxes, one's better feelings shrink from suggesting more than early closing for the fowls. A chain laid in front of the houses is said to frighten foxes away. It is well to be sure that the fowls are shut inside their house, not outside, as an error on the part of one of my lads for a single night lost me an invaluable breeding-pen which had been carelessly left *outside* when the house was fastened up.

I may mention here a curious occurrence which took place a few years ago. One of my lads, passing round a garden wall which separates two parts of my garden, was surprised to find a pitched battle in progress between a weasel and a rat. The combatants were so intent on their own affairs that they took no notice whatever of his approach. The lad, who was a first-rate shot with a stone, picked up a half brick and threw it hard at the furious little animals, with the curious result of killing *both* on the spot. His triumphal return with the two corpses may be imagined.

On another occasion, some friends shooting near my house came upon a pitched battle between a weasel and a common grass snake; one fired, intending to kill the weasel, but missed it, and killed the snake. The weasel disappeared before a second shot could be fired.

Chapter 14
Diseases

It is impossible in a small space to go fully into the many diseases which may befall the poultry fancier's live stock. It must suffice here to take a few of his probable worries and suggest how he may best deal with them.

Apoplexy

Probably caused by high feeding and want of exercise. Aperients, such as a teaspoonful of castor oil, or a little Epsom salts dissolved in water, should be given at once. If bled (from vein under wing) at once it may bring a return of consciousness. Excitement also causes it, as when a bird is washed for show, or returns from a show and is put down in his run. It is wise to see whether other birds are too fat, when a case has occurred.

Bumblefoot

Most common among Dorkings–not infrequent among other fowls. Some think it connected with the fifth toe question. It occurs occasionally in other heavy breeds, dropping from perch being a probable cause. It is a corn or abscess under the foot. If a corn, cut, or dress with lunar-caustic, or acetic acid. If it develops into an abscess, open, clean thoroughly, dress with an antiseptic lotion, and tie up till healed to keep dirt out. If cheesy-matter be found, it should be pressed out and treated as above suggested.

Canker (or Ulceration)

Usually about the head. Chlorinated soda or fluid carbolate are recommended. If in the ear I should try lead lotion, which does wonders with dogs. It often begins with discharge from the eyes, which, at first watery becomes thicker and offensive in smell. A wound in the mouth seems sometimes to lead to it.

Cold

A simple cold should be treated as human experiences suggest. A warm corner, free from draughts; light, nourishing soft food, a little tincture of aconite in the water–three drops of No. 1 tincture in half–pint of water is suggested by one authority. This cold may go on to *roup*, a very dangerous and contagious disease (see *Roup*), so isolation at once is wise for even a cold.

Cholera

At first diarrhoea and extreme thirst. Discharges at first greenish, then thin and whitish, as in the human subject. Isolate at once. See that the other birds have shade; that the water is fresh and not sun-warmed, and the vessels clean and disinfected; and have fresh green food daily for the birds. Probably death will be too quick for your treatment: small doses every three hours of the cholera mixture sold by chemists, and a teaspoonful of brandy, mixed with fairly strong coffee–one part brandy to two of coffee–given between the medicine doses may be of some use. M. Pasteur invented, I believe, a successful vaccine for chicken cholera.

Cramp

Remove the patient to a dry, warm place, boarded floor best. Stand the legs in warm mustard and water for a few minutes at a time. Small doses of opium (quarter of a grain for chicks three months old) twice or thrice daily are recommended.

Cropbound

If possible knead the crop; if no good effect, pour warm water down the throat and try again. If possible, empty the crop through the mouth: then give a dessertspoonful of castor oil. Feed moderately for some days, and on soft food only; a second small dose of castor oil in two or three days may be advisable. A tonic is good, and the bird must be kept fairly strong to enable the crop to contract, or it dangles in an unsightly manner, and the bird can scarcely ever really empty its crop thoroughly. In case the crop cannot be emptied by the kneading, it may be necessary to cut it. Avoid large blood vessels, make a cut big enough to admit a finger or teaspoon, empty the crop thoroughly, taking care to leave no obstruction to still prevent the passage on ward of food from the crop. A cropful of grass is sometimes the cause; often merely over-eating of hard corn, which, when water is drunk, swells, and blocks the crop. To stitch up the crop use a fine needle and horsehair or fine silk. Take care first to stitch together the inner skin of the crop, then the outer. Many people find no difficulty in doing this, or even in cutting out a diamond-shaped-piece from the crop before sewing, in order to contract it. This is useful when the crop is permanently loose.

Debility

For a valuable bird, raw eggs put down the throat twice or thrice daily will be no extravagance. Parrish's Chemical Food and cod liver oil, equal parts–a teaspoonful of the mixture given twice a day–is excellent. Cooked lean meat is also useful.

Diarrhoea

I usually give for this a little arrowroot or ground rice mixed with water into a large pill, and put down the throat. Feed on boiled rice, and if necessary add powdered chalk to the food. One prescription given by an old breeder is–5 grains chalk; 5 grains rhubarb; 3 grains cayenne pepper made into a pill, and in very severe cases half a grain of opium added to the pill. If the diarrhoea goes on to dysentery (that is, passing of blood with the discharges) I recommend a merciful and immediate death, and burial with disinfectants. Chlorodyne, 2 to 6 drops in a teaspoonful of warm water seems, in cases of diarrhoea, likely to give good results.

Egg-Bound

In mild cases a dose of castor oil or Epsom salts answers the purpose. If that does not answer shortly, chop some groundsel, make it with butter into a big pill, and administer. A feather dipped in sweet or linseed oil and passed up the vent and round the fore part of the egg often succeeds. Great care in handling must be exercised, to avoid breaking the egg inside the hen. Olive oil injected with a small syringe is a further method. The bird should be kept very quiet for a day or two, and fed on soft, non-stimulating food.

ELEPHANTIASIS

See *Scaly Legs*.

ERUPTIONS

See *White Comb*.

EGG-EATING

This is rather a vice than a disease, and very troublesome to cure. The only successes I have obtained were by leaving rotten eggs, which had been well sat on, about at the disposal of the egg-eater. It has been most amusing to see the disgust of the offender, after a greedy rush at a rotten egg. If the peck explodes the egg, so much the better. I have tried the plan often recommended of filling egg-shells with cayenne and mustard, but some hens seem to like it; others merely leave it. The rotten egg plan is more successful according to my experience. If the egg-eater is very valuable and all plans fail, a nest box contrived so that a new-laid egg at once rolls under a sheltering cover is good. The difficulty is that all hens won't use such elaborate nest boxes. A sight of the disappointed hen looking for the missing egg is some consolation for annoyance and reward for trouble.

FEATHER-EATING

Some birds will eat the feathers from their companions, and even their own feathers. From this they sometimes go on to eating the flesh of their neighbours, and I have known the stern of a hen covered with wounds from the vicious pecks of her sister. The plans which have succeeded in stopping it are various and opposite; giving raw meat, giving no meat, aperients, tonics, morphia, etc. Lettuces in quantity do good, and Mr. Lewis Wright's prescription of morphia was suggested by the sedative nature of lettuces, one-eighth to one-fourth grain daily of acetate of morphia, a grain of calomel twice a week, a little carbonate of potash to the drinking water; these he finds often efficacious. A little flowers of sulphur in the soft food twice a week often has good effect. A large run in a field, so that the birds scatter; a heap of ashes to scratch in and get occupation; covering the parts attacked with a wash of bitter aloes; these are all worth trying. Unless in the case of a very good bird, the motto should be rapid success in the treatment, or death to the culprit, as one will soon teach a whole pen the habit.

Fits

When I have a bird in a fit; which rarely happens–except occasionally, while it is being washed for show–it is at once put under the tap, and the cold water allowed to run just over its head, which is held in position for the purpose. This dripping of cold water on the back of the head usually brings a return of consciousness, when the bird is put away in a quiet pen, or in a hamper to recover. If the water treatment fails, the only likely plan would be to bleed it by opening the vein under the wing.

Frostbite

It sometimes happens that cocks of the large-combed varieties get their combs and wattles frostbitten. In colder countries it is a common occurrence, but even here in England I have had a case (a slight one) in a Brahma. The treatment is like the treatment of a frozen nose or face in Russia and other cold countries, namely, to rub the parts affected vigorously with snow or apply very cold water, rubbing all the time. When cured, glycerine or vaseline may be applied. If necessary, oiling the combs and wattles in the morning is a preventive, as water can't settle and freeze on wattles when the bird drinks if they are oiled.

Gapes

This curious disease is caused by the presence of small white worms in the windpipe of chickens. They are parasites, as to whose methods of arriving at their destination some discussion has arisen. There seems no doubt that they are connected with the presence of a parasite louse or tick to be found on the heads of young chicks. This view is borne out by the experience of St. Bernard breeders, who are especially troubled with worms (threadworms) in puppies. I have found 200 in a six weeks' pup. These worms are undoubtedly connected with a form of louse to be found on dogs. The fact is that in some way the louse acts as the carrier of the eggs or young of these worms, and deposits them in the nostril or mouth of the unfortunate recipient. I have had much trouble with the dog form of this disease, but scarcely any with gapes. A feather dipped in something injurious to the worms but not to the fowl, pushed down the throat, and twisted round so as to draw up the worms with it, is a good plan. I should, if I had occasion, try my feathers dipped in Jeyes' Fluid 1 part, water 60 parts, of which I have a very high opinion as a harmless and powerful disinfectant and insect killer. Holding the chicken over carbolic fumes (burning a little carbolic acid in an iron spoon over a candle) is recommended also. Some dressing to the head of the chicken to destroy the lice is excellent in an early stage. I should try paraffin oil 1 part, sweet or castor oil 2 parts, well mixed, with a little flowers of sulphur stirred in to the consistency of thin cream. This dabbed on the heads of the chicks, and pyrethrum powder well rubbed in all over their bodies, would probably meet the case.

Giddiness

If a fowl staggers about the run, or runs round in a circle, in a sudden and meaningless way, proceed at once as advised under *Fits*. It may go on to the apoplexy stage.

Gout

Legs and feet inflamed, swollen, and hot. Aperients at first; say half a compound blue pill at once, and the other half two nights afterwards. If necessary, Epsom salts to keep the bowels open, Keep the bird dry; give plenty of green food. A pill of half grain extract of colchicum twice a day, after the aperient has acted, is recommended on good authority.

Indigestion

Often caused by want of proper grit and gravel to aid digestion. Too much soft food and too little hard corn also weakens the action of the gizzard–or the bird may be generally "below the mark," or the liver sluggish, or have had too little exercise. In most cases of the sort great thirst is present, and fowls will fill their crops with water, sometimes to such an extent that the crop never properly contracts again. Give, therefore, only a moderate amount of water after each meal, leaving none for the bird to drink freely from. Empty the crop, and give first half a compound rhubarb pill and half a compound blue pill, together–the ordinary pills as sold by chemists. Feed moderately on soft food only, and a *little* lean meat, cooked and chopped fine. After a day or two give tonic, half teaspoonful each Parrish's Chemical Food and cod liver oil twice a day. A digestive pill, such as is prescribed for humans, cut in half and given twice a day is good. It should contain rhubarb, pepsine porci, aloes, and nux vomica. Most chemists keep such in stock. Exercise and plenty of grit and gravel are the main things to remember for future treatment.

Inflammation of Rump Gland

This gland is placed just above root of tail, and contains the oily secretion which fowls use for anointing their feathers. If this gland is obstructed, inflammation will probably ensue. Open with needle, squeeze out any matter, and foment with warm water. If the matter again collects it must be kept open, squeezed out if necessary, till it returns to a normal condition.

Leg-Weakness

Young birds, especially the cockerels of the heavy breeds, such as Brahmas and Cochins, often outgrow their strength; their bodies get too heavy for their legs, they squat about on their hocks, and in the end become in-kneed or bow-legged. Insufficient exercise is one cause; bad feeding another; constitutional weakness another; want of bone-making material, such as lime and gravel, another. Give bone-dust in food; one teaspoonful of Parrish's Chemical Food, and of cod liver oil, mixed in equal proportions, twice or thrice a day; nourishing, not fattening, food; a little lean meat; and encourage to take plenty of exercise. Rubbing the joints and legs with some stimulating embrocation is the best external plan. For a medical change, good authorities prescribe sulphate of iron, one grain; strychnine, one-twentieth grain; phosphate of lime; five grains; sulphate of quinine, one grain; make into a pill, give three times a day, early, late, and halfway between. This is an excellent prescription for the purpose.

Insects

Fowls are tormented with fleas, lice, and ticks. Frequent powdering of the birds with pyrethrum powder, Jeyes' powder sprinkled on the ground of the sleeping places, lime-washing the houses and perches, especially any chinks or cracks, with ordinary limewash, to which a little size and one part in sixty of Jeyes' Fluid have been added, will greatly help to keep these pests down. Painting the perches with one part of paraffin and three parts water is also excellent. Good heaps of sand or ashes for dusting help the birds to clean themselves. A little sweet oil rubbed on the back of the head is also good. Insects, or their eggs, are often brought in with the straw. Old straw should be buried or burnt, and cleanliness generally attended to.

Liver Disease

The one great trouble to the breeder of Asiatics. Masses of tubercules may be found in the liver, and at times in many of the organs of birds badly affected. Damp is one of the causes of this trouble, but it is also constitutional, and probably is communicated from one bird to another by the droppings. Mere indigestion and torpid liver in the early stages may be dealt with by podophyllin or blue pill, or even Epsom salts. When the droppings have been of a yellow colour for some time, and mild measures have proved unavailing, the best plan is to kill the birds, as recovery from a bad case is, I believe, impossible; and a lingering death), after poisoning the ground for others, is the only prospect for the invalid.

Pip

Strictly speaking, like the celebrated "Mrs. Harris" of Dickens, "there is no such person." Indigestion and inflammation of the throat produce a foul, dry tongue. Castor oil should be given; careful feeding, a warm place, and careful nursing are the best remedies. Chlorinated soda applied to the parts affected is said to be of benefit.

Moulting

Though not a disease in itself, it often leads to weakness and illness. If the fowls can't get rid of their feathers they should be helped, the dead feather being pulled out to make room for the new ones. The head feathers seem often a special difficulty, and a well-known breeder once told me how he lost a valuable cock because he could not moult his head feathers! When I replied, "Why did you not pull them out?" he answered, "I never thought of it!" Tonics, such as the Parrish and cod liver oil recommended elsewhere for indigestion, or a little iron in the water, nourishing food, and plenty of green food are the best helps. If birds get very thin in moult, an extra meal or a tonic is advisable, as the strain on the system of throwing off the old feathers and growing the new is very heavy. A little extra cooked meat daily during moult is of great service, and if during the latter part of the moulting stage the new feathers seem backward in growing it is well to mix a *little* flowers of sulphur with the soft food about twice a week. Sulphur is no doubt of great assistance for producing the new feathers.

Rheumatism

See *Cramp.*

Roup

The most dangerous and contagious disease to which poultry is liable, though personally I have had little experience of it. Beginning with a cold, it goes on to a thick discharge from nostril and eye, and even forms up into a sort of cheesy mass. Isolation is absolutely necessary at once, as these discharges drop about, especially into the drinking water, and the disease may go through a whole yard. Fresh air in the houses without draughts is the best preventative. Carefully disinfect the drinking-vessels which roupy birds have had access to. Give the isolated bird a dessertspoonful of castor oil, bathe the face and postrils with disinfectants, such as diluted Candy's fluid; Labarrague's solution of chlorinated soda, one part, water two parts, is also recommended; and after improvement, tonics should be tried, such as Parrish's Food and cod liver oil. If no improvement in a few days it is best to kill the bird.

Scaly Legs

These are caused by a minute insect, burrowing among the scales of the leg, and building up a sort of coral formation of its own. It. is not strictly a disease, and is easy to cure. Birds running in damp grass develop it more rapidly; it is no doubt very contagious, and sometimes grows to a most unsightly extent. Wash the legs as well as may be, steeping them in warm water, and brushing off all that will come away with a nail brush. Then dress well with paraffin oil, in which a little flower of sulphur has been mixed. Repeat every other day till cured.

Soft Eggs

The production of soft eggs–that is, of eggs without the proper firm shell–is usually the result of one of two causes, or of both combined. The first is an irritated or inflamed condition of the ovaries, probably the result of too stimulating food, or of the bird being overfed and too fat. The second is the want of lime, grit, etc., for the formation of the shell. The best line to take is, firstly to give a dose of Epsom salts, then stop all stimulating food, and give plenty of lime-making material for the hen to eat. Old mortar is excellent; also get some oyster shells, bake them well in the oven, when they will be easier to break up with a hammer. The bird will eat this herself, but a little of the finer mixed with the soft food is desirable, and is good at all times. The hen should be isolated for a few days in a quiet pen.

White on Face or Comb

This is usually an indication that the bird is out of sorts. It sometimes extends to the neck feathers. Cleanliness and plenty of green food are essentials. Dress parts with carbolic oil, or an ointment of cocoanut oil, one ounce, powdered turmeric one quarter of an ounce.

Wrytail

This is often a deformity, and very hereditary; but often it is a sort of trick, acquired when the bird has in its young days outgrown its strength. I have known cockerels acquire it, yet quite lose the habit when full-grown adults. If a mere trick, some benefit may result from penning the bird, and turning it round the pen several times a day in the direction which compels it to carry its tail on the opposite side. In most cases of wrytail the tail is carried to the left, a curious fact. A bad case is only to be cured by slaughter, when cooking usually follows.

Chapter 15
Description of Various Breeds

Brahmas

This fine breed is said to have been originally brought from India, from the river in which the name Brahmapootra is derived. A pair were taken thence by a sailor to America, these probably crossed with some Shanghai or Chinese variety, and eventually came to this country.

In 1852, light Brahmas of some sort were presented to the Queen by an enterprising and advertising American named Burnham. Brahmas had been previously exhibited by Dr. Bennett in 1850, at Pittsburg. About 1853 or 1854, dark Brahmas began to be known in this country, and a year or two later, my own father had some, and bred them so well as to win on one occasion first, second, and third prizes, under the late Mr. Hewitt. That originally the darks and the lights sprang from a common source is certain; in the earlier days the peculiarities of colour were not so distinct, and it was quite possible to breed gradually lighter or darker birds so as to get either lights or darks from a common stock in a season or two. Careful breeding for one or the other has greatly altered this state of things, and no one would now try to breed darks from lights or lights from darks. In fact, in the dark variety, the anxiety to get black-breasted cockerels and beautifully marked pullets has led to a further practical subdivision of darks into cock-breeding strain and pullet breeding. The cock breeders are the show cocks, and their mates the dimly marked, in fact often quite unpencilled, hens, fine in general quality, bred of first-rate cock blood, but not likely, from any pencilling of distinctness, to throw mottle-breasted and fluffed cockerels.

The pullet-breeding cockerels and cocks are as a rule finely laced all over breast and fluff, though some cockerels, beautifully laced as young birds, when they moult out as adult cocks, come practically black breasted. These are often most valuable for pullet-breeding.

The pullet-breeding pullets and hens are simply the most perfectly marked birds, both as to evenness of pattern and completeness of distribution of pattern all over the bird, fineness of the bands of pencilling, and clearness of ground colour; this latter should be as silvery a blue as possible, quite free from any brown or red shade.

The Brahma should be a very large bird, with a remarkably small head in proportion. The comb should be what is called "a pea comb": namely, should have three parallel ridges, the centre one being slightly higher as a rule than the two outside ones. It should fit closely to the head, not shaking at all when the bird moves, and is best when at the back it still continues to cling closely to the head. Some combs rise to a peak at the back. These are now seldom seen in a show pen. Single combs were occasionally seen some years ago, thanks probably to a cross, but they are practically extinct in this country. One curious characteristic–and very characteristic it is, and not enough sought after by breeders–is the peculiarity that the deaf ears should fall below the wattles. Nothing injures the fine-bred look that a Brahma should have more than coarse, heavy wattles hanging about under his head.

The lights should be of the purest *blue*-white, not yellow or creamy; the neck hackle white at the edges, but heavily and distinctly striped with intense black, the black striping to run as far in towards the neck as possible.

DARK BRAHMA COCKEREL
Winner of 15-Guinea Challenge Cup, Crystal Palace, 1901
(Property of G.W. Henshall, Esq.)

The body, wing, and saddle should be covered with plenty of pure white feathers, which should not be soft and fluffy like the Cochins, but rather hard and close-fitting. The legs, should be well feathered, the feathers standing out from the shank; especially important is it that there should be no failure of feather *at the difficult point,* that is just under the hock. The foot must be well covered with feathers standing well out, and especially must the middle toe be well feathered. A certain amount of black may be allowed in the foot-feather of alight Brahma, and it must be kept in mind that a Brahma, whether dark or light, is strictly "a study in black and white." The tail of the Brahma should be carried nearly upright, the saddle and cushion rising gradually up to the tail. In light and also in dark cockerels, the tail should, if possible, be black laced with a fine white edging. In old cocks this sometimes moults in excess, and is thought to be "white in tail," but if confined to the edges of the feather is scarcely a fault, even if excessive in quantity. White in tail in an unmoulted cockerel is a far more serious fault. In all Brahmas a green reflection or sheen is to be preferred to a purple one, but purple reflections are not to be considered fault in the show pen.

The dark cock, for show purposes, should be black breasted and fluffed, or *slightly* and evenly laced or spotted all over the breast and fluff. This lacing must not take the form of irregular splashes of white. The hackle striped, as in the lights, but not quite so heavily for preference; the saddle hackle also heavily and soundly striped, and the tail coverts should be especially brilliant. This saddle striping is often seen in lights, and some breeders like it for certain breeding purposes, but the tendency of opinion is clearly to the effect that a light Brahma should have a clear saddle.

The tail of a Brahma cock, whether light or dark, should at its best be what is called a "black-cock tail." The effect is that the two or four upper sickle feathers turn apart from one another. It is very beautiful when seen in perfection, and rarely met with in a bird which is not full of Brahma character. There are two more or less divergent types of Brahmas: the long necked and moderately long legged, and the short necked squat Cochiny type. The former is decidedly the better both for health, appearance, fineness of flesh, egg production, and true Brahma character. There should be, except in shape of head–which should be short–a certain amount of "gamey" appearance about a Brahma cock; at any rate, a look of activity and alertness. The general characteristics of the dark cock are similar to those of the light cock given above.

The hens must not be too leggy. Leggy cockerels, however (and occasionally leggy pullets), often settle down into the finest birds when they moult out as adults.

The dark Brahma pullet or hen requires special description as to her markings, though in shape, feather, and general character, she should be similar to her light cousin.

The feathers of a dark Brahma pullet or hen should be of a light or medium grey (grey-white mixed with black) ground, sharply pencilled with as many bands as possible of narrow black pencilling.

The outside band should be fairly near to the edge of the feather, or it gives, when the feathers are lying closely one over another, the impression of want of pencilling, though the feathers, except just at the edges, may be full of pencilling. I have seen eleven bands of pencilling on many feathers. The shape of the pattern varies in various strains.

Some judges do not like much pencilling in the *hackle* of a show hen or pullet, but therein I differ from them, and in a breeding bird I value it highly. Therefore from a "show" point of view it will be mainly a matter of personal taste on the part of the judge. If this pencilling can be got also on fluff, leg feather, foot feather, at back under tail, and as a pencilled edging to the tail feathers, so much the better. It is useless to mate such birds with show cocks. The pencilling, in nine cases out of ten, will be ruined. They should be mated with a cock of pure pullet blood of similar strain to themselves, with sound hackles, and evenly laced breast, fluff, leg, and foot; if the pencilling is to be found all over the shoulders and wing, and especially as an edging to the feathers between the shoulders and on the forepart of the back, so much the better. A narrow black fringe to the neck-hackle of a pullet-breeding cockerel is not a fault but an advantage.

The principles here stated for pullet breeding apply also in the similar case of breeding partridge Cochin pullets for the show pen, and will be referred to again under the head of Partridge Cochins.

Cochins

This breed may be considered to have originated the mania for fancy poultry, and to be the father of poultry shows. It is very amusing to look back to the old numbers of *Punch*, and see the pictures of the period: fine ladies taking out an enormously leggy, and to modern taste very uncochiny-Cochin cock out in leading strings for a walk in the park. Cochins were believed by some to lay several eggs a day; and many other wondrous stories of them were reported at the time. The breed did not really come from Cochin, but from Shanghai, and the birds were called Shanghais by many, especially in America, for some years. Indeed, when Brahmas were first introduced, many insisted on calling them also Grey Shanghais. However correct the name Shanghai for the now-called Cochin may have been, it has died out, and nowadays one hears of Cochins and Brahmas, but never of Shanghais. In those early days the most successful breeder was Mr. Sturgeon, with whom I often have a chat on poultry and other matters, and who has frequently been the only magistrate with myself at the Petty Sessions at Grays. It is odd that on these occasions the Bench should consist of the father of the poultry fancy (if I may so call him) and the president of the Poultry Club. When Mr. Seabrooke, a vice-president of the Poultry Club, is also present, the Bench is curiously "chickeny."

The earliest so-called Cochins, no doubt, had some Malay blood, and were not true Cochins. About 1849 some birds were sent to Mr. Charles Sturgeon of a strange sort. He killed two, and sent them to his brother, Mr. Alfred Sturgeon, to see how they were liked for table. One of them was a pure lemon bird. Mr. A. Sturgeon at once galloped over to save the rest. He found that a third had been killed, and that five were still alive; he bred from them, and one, a partridge-coloured bird, and its partridge stock, were given to Mr. Punchard, an old friend of the Sturgeon family. In 1851, or thereabout, Mr. A. Sturgeon at Birmingham, won every prize in each class for Buffs, and Mr. Punchard every prize for Partridge Cochins, with the produce of that one consignment. When, later on, Mr. Sturgeon sold off by auction most of his remaining stock, the price realised, after all the expenses of the auction were paid, amounted to £609, a nice little sale for one day! Later on Mr. Sturgeon had to buy back a bird he had sold for £5, and to give forty-three guineas for it! In the first year of breeding sixty chickens were killed by greyhounds, a loss which can be scarcely estimated nowadays.

The original birds were alike in general type, but various in colour. The five birds left alive included a reddish buff cock, a reddish cinnamon hen, and a partridge or flame-coloured hen, with pencilled feathers. The other two hens died. Then a fox took the cock, and one of the two remaining hens died, so the result of the first importation was in the end one hen and eight or nine chickens. From these really arose the poultry fancy on a big scale.

The Cochin, of whatever, variety, should be a large and very fluffy bird, standing wide, very broad across the stern, when viewed from behind, short in the leg, and very small in tail. The comb single, perfectly upright, of as fine texture as possible, as should also be the face, wattles, and deaf ears; the comb should also be quite straight, and free from side sprigs. Head very small, curved forward, neck short and full in hackle, shoulders very broad, as should be the saddle. The breast-bone not let down so deeply as in the case of the

Brahma, but prominent. The back very short, saddle rising to the small soft tail, wings well clipped up, carriage very dignified and leisurely. The legs heavily feathered down to middle toe, with soft hock feather as abundant as possible.

The hen should have a cushion falling towards the tail, with a mass of fluff in the fore-part of it, resembling a ball; the fluff also should be excessive, so that the bird should seem almost to have three balls of fluff on her, one on the back and one at each side.

Buff Cochins

Buff cochins should be either a lemon or a rich buff. Formerly lemons were preferred, but the taste of the present day seems to run to the deeper colour. The tail should be rather richer colour than the body; black or white feathers a great fault. The hackles of the cock are a richer, deeper colour than the body, fluff, and legs, which should be all as even in shade as possible. If the hen is even in shade all over so much the better. No doubt rain and weather tend to injure the evenness, especially if the feathers are exposed to these influences while still growing. In mating, choose hens with breasts of the same shade of colour as that of the cock to be mated with them.

In many of the best, the colour of the hackles so closely approaches the colour of the body in tone, that it is difficult to distinguish where the one class of feather ends and the other begins. White or mealy flights in the wings are a great defect, and very undesirable for breeding from. A tricolour bird is also much disliked, and some very rich coloured birds are much lessened in value by this defect, which quite destroys the evenness of colour all over which is so much desired. In fact it would be most unwise to breed at all from an uneven-coloured cock.

Partridge (or Grouse) Cochins

The Partridge cock is one of the grandest-coloured birds in the poultry world. The hen resembles in marking, though not in colour, the dark Brahma, and many of the best Partridge Cochin hens no doubt owe their pencilling to a not distant Brahma cross. In general points they should resemble Buffs. The cock in colour is hard to describe.

The head and hackles should be a rich reddish orange, the hackles, both neck and saddle, being densely striped with black, as in the dark Brahma cock. The back and shoulders a rich dark red, the colour of which is most brilliant. The wing bar black, showing green or purple reflections; the secondary wing feathers, those which are ordinarily visible, should show a rich bay colour, edged black on the part of the feather which is tucked away. The breast, fluff, thighs, feathers, and tail should be glossy black; and white in tail or legs, and brown on breast or fluff is very objectionable.

The hen is of a brownish colour, the shade varying in individuals, but must in each case be even all over the bird. The hackle is more of a golden tinge, but I like to see the hackle and body colour as much alike in shade as possible. The body of the hen should be pencilled all over with a narrow pattern, just as in the case of the dark Brahma hen, and as free from streak in the feathers as possible. The body colour should be brown, not yellow, and the pencilling should run well up to the top of the throat, right back on the cushion, and on fluff, legs, and feet if possible.

There is a tendency for the colour to run lighter and more yellow in tinge towards the throat; it is much disliked, and is never seen in a good show bird. This defect is known as "clay throat."

This breed, like the dark Brahma, requires to have separate pens mated up for breeding cockerels and pullets.

To mate with the show cock a hen of good cock strain with little or no pencilling, either on breast or fluff, should be chosen, and if she is too dark in colour for a show hen she will probably breed all the better.

To mate with a show hen I should select a bird of good pullet strain, of sound colour above, and of sound, rather narrow though intense black stripings in the hackle. If the breast, fluff, etc., which should be black in a show cock, are sharply edged with the brown which is so great a fault in a show cock, so much the better. In fact, as in the case of a dark Brahma pullet-breeding cockerel, I should like to have as much tendency to pencilling as possible visible all over him, provided it is confined to the edge of the feather. If it resembles small spots it is not a disadvantage.

White Cochins

These birds look beautiful on a good grass run in the country, but are not suited for town breeders, as their great beauty, pure blue white, soon gets spoilt by the smoke and dust of town life. Like other white birds, they do not all moult out equally pure in colour as adults. Those which do are of great value for breeding. The purity of colour is most necessary on the cock's side, though for thorough success it is essential to have the hens also of the purest white. Plenty of shelter is required to prevent the birds from tanning by exposure to sun and rain, and they must not have any yellow gravel or sand to dust in, which is found to stain the feathers. A little lime mixed with the dust bath is recommended by some, but I should not myself care to use it. The characteristics should be the same as those of Buffs. In breeding Whites especial care should be taken to breed from cocks with really small Cochin tails. A big-tailed Cochin loses one of the most characteristic points in all Cochins. The eye should be red; the feather as heavy as possible, with as little vulture hock as may be attainable. These birds should always be carefully washed before they are shown. When quite clean and thoroughly webbed out they are very beautiful, and look bigger than they really are in consequence of their fluffiness.

Black Cochins

These handsome birds, which were excellent in the early days of the Cochin mania, nearly died out, but have recently been revived, and now at the best shows, especially at Birmingham, are shown of such quality as any breeder might be proud to own. There is no doubt that many blacks were bred from a cross of buffs and whites, and black and white being interchangeable colours, these blacks recrossed with whites and blacks gave good results, so far as Cochin characteristics were concerned. The traces of the buff blood appeared in red or golden feathers, and some blacks moulted out with white edges to their feathers, showing thus, no doubt, their white Cochin blood. But a cross with a black bird was made many years ago, which had also a black pupil to the eye. These were birds of brilliant gloss, wanting in feather and cushion, and generally having black legs. I have no doubt that these were much the same thing as the earlier Langshans, and were probably imported as black Cochins. From this cross the colour was much improved, and the birds so bred moulted out much better. The drawback of course was the want of feather and Cochin character, but careful breeding has done wonders in these respects. Except in colour, the black Cochin should resemble the best buff or white in general characteristics. Care must be taken to keep the texture of the comb, face, deaf-ears, and wattles as fine in texture as possible. The eye and leg in blacks is often much darker than in the other breeds of Cochins.

Cuckoo Cochins

Cuckoo Cochins are very scarce nowadays, though good birds were occasionally shown some twenty-five years ago, but only by a few breeders. They should, of course, be as thoroughly Cochin in type as possible, though as a rule they were not so typical as the buffs. If any breeder cared to revive them–and they are not to be found ready made–I should be inclined to recommend a good Dominique cock, and black and white Cochin hens, especially the former. Also a pen of white Cochin hens mated with a black cock. From these two sources I fancy a good strain of Cuckoo Cochins might soon be produced, but it would be advisable that several breeders should start together, so that when they had produced Cuckoo Cochins they might have Cuckoo crosses available, and not be compelled either to lose ground again by crossing back to the breeds which had been used in the manufacture or to find the strain become unfertile through excessive in breeding.

For the description of cuckoo marking see under Cuckoo Dorkings; of course the leg, shank, and toe feathering of Cuckoo Cochins should also be barred all over to match the body marking.

Dorkings

The general characteristics of the Dorking should be the large, deep square body, the breastbone being long and prominent, white skin, white legs and feet, five toes, the fifth being clearly separated from the fourth and turning upwards, and white toenails. The fourth and fifth toes, the back ones, should be set in low. The head large, wattles large and pendulous in the cock, not so large in the hen, and rounded. Eye full and bright; comb single or rose in coloured Dorkings; single in silver greys (though fine rose silver greys, have been shown); rose in white or cuckoo.

The single comb should be upright in the cock; in the hen it should fall on one side of the head. In the cock it should be thick, firm on the head, evenly serrated, free from side sprigs, semi-circular in outline seen from the side, and of fine texture.

The rose comb should be moderately large, firm, even in its spikes, and with no hollow in the centre, square in front and with a straight spike at the back. Neck thick; full hackled, and short. Back broad and flat at shoulders, rather long, slightly narrowing towards stern. Wings carried well up, and rather large. Tail large, feathers broad, carried high, but not so much as to be squirrel tailed. Thighs stout and covered by the plumage. Legs short, stout, and spurs carried inside (Dorking hens and even pullets frequently have spurs of considerable size). The general appearance should be massive and dignified.

Coloured (or Dark) Dorkings

Colour somewhat varies, arid is not of first importance, within certain limits, as the breed is essentially a table fowl. Comb, face,

DARK DORKING COCKEREL
Winner of 25-Guinea Challenge Cup, Crystal Palace, 1901
(Property of T. Hulse, Junr. Esq.)

etc., red. Eye, orange or dark red. Beak, horn colour. The ear lobe is sometimes tinged with white, but red is far better.

Cock

Head and neck, greyish white or steel grey, striped black. Back, shoulder coverts, and saddle, white, with black or steel-grey markings. Wing-bow, mixed black and white. Wing-bar, black. Wing-bay, white. Tail, black; a little white is not a disqualification, but all black is preferable. Tail coverts, black or grey. Breast, underparts, and thighs, black.

Hen

Head, black on top, grey under eye. Neck-hackle, black, dark, or striped dark on lighter ground. Back, dark, showing white shafts. Wing-bow, similar but free from red tinge. The visible wing feathers (the secondaries) are black, shading lighter outside. Tail and tail coverts, black or greyish brown. Breast, various shades of brown are permissible, sometimes a brownish salmon tinge shaded deeper under throat. Sometimes a black spangling is seen. Underparts, similar, often running darker, and even black, on thighs.

Silver-Grey Dorkings

Face, etc., red; eye, orange; beak, horn or white.

Cock

Head, neck, and saddle, pure silvery white, the hackle as free from stripings as possible. Shoulder, white; wing-bar, greenish-black; wing-bay, white; tail, sickles, and coverts, glossy green-black. Breast, underparts, and thighs, black.

Hen

Head and hackle, silvery white; head, slight grey marking; hackle, clear black stripings. Back and body, generally, silver-grey, minutely pencilled darker, and showing white shafts. Breast, robin or salmon-red, shading to grey below. Underparts and thighs, grey. Wing-bar, grey; tail, grey, darker than body; quill feathers, black.

White Dorkings (Both Sexes)

Face, etc., red; beak, white eye, orange. Plumage, pure white. These birds look very beautiful on dry grass runs, and breed very true.

Cuckoo Dorkings (Both Sexes)

Face, etc., red; eye, orange; beak, white or horn-coloured. The whole body, hackles, and tail should be a light bluish-grey in ground-colour, each feather barred across with darker blue-grey bars, sometimes almost black. The bars should not be sharply defined at the edges, like the pencilling of Hamburghs, but should shade off into the lighter tint of the ground-colour, and the bars of pencilling are broader and coarser than in pencilled Hamburghs. The main difficulties in breeding are the occurrence of reddish or golden hackles in the neck or saddle hackle of the cocks, and also the tendency to white in tail. The colour and character is similar to those of both Scotch Greys and Dominiques, but these breeds have only four toes, and in addition, Dominiques have yellow legs. Cuckoo Dorkings are far from common, but are a particularly hardy sort of fowl.

Scotch Greys

A very valuable and hardy table-fowl, mostly bred in Scotland. They are among the cuckoo varieties in plumage, resembling Cuckoo Dorkings in many respects, but with four toes, and differing from Dominiques in having single combs and pinkish-white legs. Both have red ear-lobes. Good classes of Scotch Greys are occasionally seen in the north.

Game

Probably the oldest breed of fowls in existence. Cockfighting dates back beyond Greek and Roman periods, and no doubt existed in Persia and India at a very early date. Whether the Gallus Bankiva (jungle fowl of India) or a larger bird be its ancestor seems difficult to determine, but that a race selected for its pugnacity and hardness of feather has been kept for many centuries for fighting purposes in the East is certain. In England cock-fighting was long a popular sport, side by side with bearbaiting; but the law, and a better state of public opinion, which produced the law as it now stands, have to a great extent done away with what was, at its best, a cruel and debasing sport. That cockfighting goes on even now is certain, and from whispers one hears occasionally, one wonders whether the police do not sometimes wink an eye which should be turned on cockfighting and its abettors. Some complaint is made that the modern Game fowl, as seen in our shows, is but a degenerate specimen of the old sort. That it is longer in neck and leg there is no doubt; in those yards which no longer test for gameness it is likely that a less determined race will be bred than in the old times, when a bird to be selected for breeding would have to prove his right to existence in many a hard-fought main. Still, if the reasons given for the necessity of dubbing are well founded, there is still no lack of pugnacity among the modern Game. Many of the great families of England had their own strains of Game, among which those of the Earls of Derby were much noted.

General Characteristics

Head fine, long, and keen; beak stout, especially at base, slightly curved; comb, deaf-ears, wattle, fine in feature; comb, single, small, thin, upright. In most cases Gamecocks, when shown, have had comb, wattle; and deaf-ears removed or dubbed, as it is called. This is done, it is alleged, to enable the breeder to keep several cockerels running together, and this is said to be impossible when the comb-wattles, and deaf-ears are left on for the birds to catch hold of when fighting. How much truth there is in this assertion is doubted in some quarters, where it is alleged that, apart from fighting requirements, dubbing is carried on to improve, as it is thought, the appearance, and in consequence of a certain fashion in the matter.

The head should be quite free from the heavy hanging eyebrow which is seen in the Malay, and occasionally in Game fowl, thanks, no doubt, to a Malay cross. Neck, arched and rather long than short. Hackle, short and scanty. Throat and head, lean. Breast, rather full, the body tapering to a point at the root of the tail, but the breast should not be deep or let down, as in many breeds. Saddle, narrow, and the saddle-hackle short and scanty. Wings, medium or small in size, but strong and muscular, standing out at the shoulders. Thighs, muscular, round, and hard. Toes, long and straight, back toe set low, and lying almost flat. Legs, long with fine scales, not flat. Tail, small, and carried low. Whole plumage glossy and hard. High bred, elastic look, combining strength with great activity. The cock should carry the tail rather higher, but the hen not much above the horizontal.

Black-Red Game

Beak, horn, slightly yellow; face, etc., bright red; eye, fiery red. Legs, willow or olive (white or yellow legs objected to).

Cock

Head and hackle, bright orange-red, not striped. Back and shoulder, rich crimson-red, shaded orange. Saddle, orange-red. Wing-coverts, black with green reflections; secondaries, bay outside, black inside, with black ends. Tail, glossy black. Breast, throat, and underparts, glossy black. Some cocks are darker in the shade of red than these, but are not so much liked for show.

Hen

Hackle, golden, with narrow black stripes. Breast, reddish, with a salmon tinge, shading towards thighs to an ashy-brown colour; elsewhere a brownish colour, salmon on thighs, with very dim markings rather resembling those of the Partridge Cochin, but very indistinct except sometimes towards the tail. The less pencilling the better.

The Wheaten or Weedon Hen is also used for breeding Black-red cocks, also for Piles and Duckwings. The tendency to breed soft feather is considered the worst objection to these hens. They have a pale buff breast; back, reddish buff, or light fawn, with as little sign as may be of the *dim* partridge pencilling; head, golden; hackle, buff, striped sometimes with black; tail, black, often edged or stained buff; thighs, greyish; legs, willow or olive; comb, face, etc., red; eye, very bright red.

Brown-Red Game

Beak, black or blackish; face and head, purplish; eyes, large, black or nearly so; legs, black or very dark willow.

Cock

Head and hackles, from deep orange-red to brighter shades, as lemon or golden. Hackle, in the starling-breasted cock may be more streaked than in the laced. Back and shoulders, golden-red; saddle, rather lighter, running to lemon at the sides, well striped. Breast, a shade of black edged or laced with light bay or lemon, the light colour not running into the feather. Tail, glossy, greenish-black. Thighs and underparts, black.

Hen

Hackle, various shades of gold or lemon, striped with black. Body, black or blackish-brown. Breast may be laced. Tail to match body.

Duckwing Game

This richly coloured bird is mostly bred by crossing with Black-red Game, especially Duckwing cocks, the pure Duckwing hen being mated with the Black-red cock. Duckwing cocks are also crossed with Wheaten hens to produce Duckwing cockerels.

Yellow Duckwing Cock

Face, etc., bright red. Eye, red. Beak, horn colour. Legs, willow, sometimes yellow, which are not so good. Hackle, light straw (almost white on head), quite free from black stripings. Shoulders, brassy maroon, free from light streaks. Back, orange, shading off to a straw rather darker than the hackle. Shoulder-butts, black. Breast, tail, and underparts, black. Tail, greenish-black, grey at roots.

Yellow Duckwing Hen

Head, grey. Hackle, silvery grey with narrow black stripes. Breast, salmon-red to fawn; fluff, steely grey inside. Back, shoulders, wings, slaty grey, dimly marked but not pencilled; tree from any ruddy tinge. Tail, black except two top feathers, which are grey; secondaries, grey; primaries, black.

Silver Duckwings

These were a true breed of Game, and bred true to colour years ago, when mated among themselves. In some colours of Game it is necessary to cross the colours to get the desired results. The face, etc., should be red; beak, horn; eyes, red; legs, willow.

Cock

Head, neck and saddle, hackle, back and wing-bow, or shoulder, should be clear silver. Wing-bar, black; wing-bay, white. Breast, thighs, and underparts, black. Tail, black, green gloss, grey at roots; lower tail coverts, straw colour.

Hen

Head and neck, silvery grey, the hackle with narrow, dark stripes. Back, saddle, shoulder, and wing, grey. Breast and thighs, pale salmon.

Pile Game

Face, comb, etc., bright red; eye, red; beak and legs, yellow.

Cock

Head, bright orange. Neck, orange or chestnut, not striped. Saddle, rich red, darker on back. Shoulder coverts, white. Wing-bow, rich red; greater and lesser coverts forming wing-bar, white; the triangle, or wing-bay, red; wing-butts, white. Breast, thighs, underparts and tail, white.

Hen

Neck-hackle, pale gold. Breast, a salmon or chestnut shade; elsewhere white.

Black Game

Face, etc., red; eyes, red; beak and legs, black (the legs may be a *dark* willow). Plumage, black with green reflections.

White Game

Face, etc., red. Eyes, red. Beak and legs yellow. Plumage, white.

Birchen-Yellow Game

Cock

Dull-red face, red eye. Beak, horn colour. Hackles, cream. Back, cream and rusty brown. Breast, clay, mixed with brown. Legs, willow. Tail, black. Shoulder, cream, with rusty brown wing-bar and bay.

Hen

Face, etc., dull red, with blue-red deaf-ear. Upper parts dark grey, darker on back and shoulders, and sometimes black on wing-bow. Thighs, black or dark grey. Tail, black. Legs, dark willow.

Ginger-Red Game

A variety of the Brown-Reds.

Cock

Tail, reddish purple. Eyes, brown. Beak, blackish. Upper colour, rich red. Wing-bar and triangle, brownish red. Breast, ginger-red. Thighs, dusky red. Tail, black. Legs, olive or dark willow.

Hen

Comb, face, etc., purple. Head, yellowish brown, as is the general plumage. Neck-hackle, yellow with black striping. Breast shades to a darker dusky brown at lower parts.

Henny Game

These, no doubt, had been bred true for generations in Cornwall and Devonshire, and it is said that uncrossed they always breed the true Henny cocks. As a rule they have white or pinkish white legs, though some have yellow or willow. The peculiarity is that the cocks have true *hen's* plumage and hen's tails, and it was, no doubt, from some bird of this blood in part that Sir John Sebright got his idea of breeding a hen-tailed strain of Ban-tams.

Old English Game Fowl

This, the older form of the modern improved (?) Game, is no doubt unsurpassed as a bird to cross with Dorkings or Brahmas, to obtain a first-class table-fowl. As compared with the ordinary show Game-cock, the "old English" bird is chiefly remarkable for very short legs, though perhaps not short as compared with some other breeds. The comb, wattles and ear-lobes should be small and fine in texture, the beak rather curved and strong, especially at base, as it is one of the important weapons in the Game-cock's armoury. The neck to be long or reachy and strong to wield his weapon, the beak. A broad chest, the body tapering more or less like a pear towards tail. Long, powerful wings, muscular thighs, which should be short and wide apart; the shanks not flat on shin. The four toes long and well spread, the hind toe well backward and flat on ground; the spurs powerful and set low. The plumage glossy and fine, hard in handling, but the bird should feel active or corky in the hand.

The head should be of medium length, tapering; the eyes should match, the neck-hackle long and wiryl the back short, the tail hard and strong, main feathers broad and curved, and sickles abundant. The whole appearance of the bird should be springy and active, perhaps the word alert expresses its most noticeable characteristic.

The more usual colours are black-red; with Partridge hen, ginger-red, brown-red, red pile; silver duckwing, white, black, black with brassy wings, and spangled.

The club devoted to this breed also recognizes many further varieties, such as Hennies, Muffs and Tassells in all colours, Furnesses, Polecats, various Duns, Piles and Duckwings, Birchen-greys, etc.

Malays

This breed is not very popular in England, though years ago its great size recommended it for crossing to gain size. The cruel, sour expression, long leg, and ungainly appearance generally, is against popularity, but it is so distinct that some details should be given.

Head, broad and long, with heavy overhanging brows, stout, much-curved beak, and generally cruel look. Face very bare of feathers. Comb, peculiar, a small round convex lump, covered with small projections, deep red or purplish, rather like half a strawberry. Neck, very long. Hackle, hard and scanty. Shoulders, carried very high and projecting. Thigh bone very long. Eye, pearly yellow, *not red.* Body slanting downwards. Tail drooping, medium in length. The general appearance somewhat like a leggy Indian game, but much more heavy in head. The commonest colours are black-red cocks, and cinnamon or brown Wheaten hens, but most Game colours have existed.

Indian Game

This breed is no doubt originally from the Malay, probably crossed with the old English Game. It is now popular, especially in Cornwall, and large numbers are bred and shown every year. Its general appearance is thick-set and sturdy, very broad at the shoulders, with prominent shoulder-butts, the body tapering from shoulders to tail; back, fiat. The head rather long and rather thick, between the thickness of the Malay and the fineness of the old English Game; skull, broadish; neck, medium, with short hackle; beak, yellow or brown, or the two mingled, thick at base; face, red, fine in texture; throat, dotted with fine hairs; comb, pea as in the Brahma, set close and small. Breast, deep but well rounded; wings short, carried close, and rather high in front. Legs round, very strong, but without the length of thigh of the Malay; the shank moderately long, gamey but not stilty, and as rich orange in colour as possible. The four toes to be long and strong, the back toe low and almost flat on the ground. Tail, of medium size, carried rather low; hard feathered, and the secondaries and coverts short and narrow; colour of tail, green-black. The cock in colour is principally of a glossy green-black, intermingled on back and at base of hackles with rich bay or chestnut, though this colour at base of hackles is almost covered by the darker feathering. The triangle on the wing, formed by the secondaries between the wing-bar and the wing-butts, is also bright bay or chestnut.

The hell, apart from colour, should resemble the cock, except that the tail should be rather short, and though carried low, scarcely so lowas the cock's; the tail feathers should well overlap, or, to use the curious phrase of the Indian Game Club standard, should be well "venetianed," a good use of the metaphor.

The hen's head should be black like the cock's; the hackle green-black at the upper part, but with a rich chestnut centre to the feathers at the lower part. The breast and underparts, rich chestnut edged with green-black, with a distinct double lacing on the larger breast feathers; lower, and on the thighs, the marking runs more indistinct though similar in character. The back and wings similar, and double laced, the shoulders and wing-bows being especially distinct. The wing-bars laced as body, and often a little "peppered." The secondary wing feathers are black on the inside, and rich chestnut on the outside, with narrow edge of green-black.

Tail and tail coverts simiiar to rest of plumage, a green-black edging to a chestnut ground.

Disqualifications

In addition to the ordinary malformations fatal in all breeds are single or Malay combs, red hackles and bent legs.

Indian Game are by some much liked for crossing for table purposes, and it is no doubt a useful breed for the purpose, though the old English Game would probably give best results when crossed with most other breeds.

ASEELS

Another fighting breed, similar in some respects to the Indian Game, and certainly one of the most pugnacious of all breeds.

Cock

Comb, pea. Eye, bold and full, yellow or yellowish grey. Wattles, very scant. Ear-lobes, small and fine. Face-wattles and ear-lobe red. Neck, medium and arched. Beak, strong and curved, yellow, or yellow-striped horn, yellow preferred. Throat, dotted with small feathers. Neck-hackle, moderately short, just covering base of neck, greenish black, with brown-crimson shafts. Back, very broad across shoulders, narrowing from thighs to tail; sloping, in convex line from shoulder down to tail. Colour of back, glossy green-black, with some dark crimson-red intermingled. Saddle-hackle, similar in colour, but less-black. Breast, projecting slightly beyond the wings. Colour black, as are the underparts. Wings, medium, standing out at shoulders; wing-butts, black; wing-bows, black with some crimson-red; wing-bars, black; secondaries, lower web bay edge, running black towards quill; upper edge black. Thighs and shank, stout, medium, the shanks of as deep a yellow or orange as possible, and not flat. The back toe nearly flat on the ground. The general appearance is between that of the Game-cock and the Malay; the head being not so seen as the Game nor so heavy browed as the Malay, which latter is also much more leggy than the Indian Game. Tail, medium, close, drooping, green-black; secondary sickles and coverts: short and narrow.

Hen

Apart from colour, similar to the cock. Tail, rather short, carried rather low; colour, a shade of brown, from chestnut to mahogany. Head and throat, green-black. Neck-hackle, rather short, black, with narrow bay shaft. Breast, bay, each feather with double lacing of black, the inner lacing being most distinct. Underparts similar, but less distinct towards thighs and fluff. Back and shoulders, similar in colour, especially clear on wing-bows and shoulders. Tail-coverts, similar but less distinct as a rule. Secondaries, visible part like the general ground colour, with narrow lacing of greenish black. Wing-bar, laced, and often a little peppered. The ground colour must not be mealy, nor exceedingly light or dark. Single or Malay combs are fatal defects.

Hamburghs

The general appearance of the Hamburghs is one of great smartness and liveliness, but without the strut of the Spanish, the Spangled and Black being somewhat larger and more substantial than the Pencilled. Head, rather small and short. Beak, rather small, horn colour (in Blacks may also be black). Comb, rose, wide and square in front, tapering to a long spike pointing somewhat upwards; to be flat on top, full of points or "work," set firm on head. Deaf-ears, white and circular. Legs, dark leaden blue; in Blacks, blackish. Neck, rather long, arched, carried back. Hackle, very full and sweeping. Wings, large. Breast, full and prominent. Legs, short and slender; toes, slender. Tail, very large, well furnished with sickles, which are curved, carried high, but not " squirrel." Size, small to medium. Eyes, red Wings carried rather low.

Silver Spangled

Cock

Colour, pure white, marked with black spots as follows: Hackle, pure white, spotted with black at the bottom. Back, shoulder-coverts and wing-bow, white spotted, with long narrow spots. Wing-bar, pure white, with heavy round spots at the end of the feathers forming two parallel bars across the wing. Primaries, white, spangled at the ends. Secondaries, white, with large half-moon spangle at end. Saddle, white, spangled at ends. Breast, underparts and thighs, white, spangled at end of each feather with large round spots, as

SILVER SPANGLED HAMBURGH COCKEREL
Winner of Challenge Cup, Grand International Show, 1901
(Property of A. Blackley, Esq.)

large as may be, but the spangles narrowly divided by the visible white. True tail white outside, whitish inside, each feather spangled at tip. Sickles, secondary sickles, and tail-coverts, pure white, ending in heavy; black spangles.

Hen

Hackle, white, spangled black at ends. Tail, white, spangled as roundly as may be. General body, plumage white, with large round black spangles, which must be clear, not over-lapping, but showing white between each. Wing-coverts, with large spangles, forming two parallel. lines as in the cock. In other respects as the cock's colour. Evenness and uniformity of marking of great importance.

SILVER SPANGLED HAMBURGH HEN
Winner of 1st Prize, Grand International Show, 1901
(Property of A. Blackley, Esq.)

SILVER SPANGLED POLAND COCK
1st and Cup Crystal Palace, 1901
(Property of F. Anderton, Esq.)

Golden Spangled

The same as the Silver, with the following exceptions: Ground colour, rich golden bay. Neck and saddle, hackle striped with black, not spotted. Tail, black, not spangled.

Silver Pencilled

Cock

Hackles, back, shoulder-coverts and wing-bow, pure white. Wing-coverts, lower web white, upper web black, an indistinct wing-bar being seen, owing to the black of the upper web, showing just at tip on the outer web also; primaries, white outer, black, inner web; secondaries, lower, visible web white, with narrow strip of black next the feather; upper web black, having whitish or grey at upper edge. Plumage, generally white, but for slight pencilling on fluff. Tail, glossy green-black, the sickles and Coverts being edged all round with *narrow* lacing of white.

Hen

Silvery white hackle. The rest of plumage, So far as may be, pencilled across with narrow lines of greenish black, the lines of pencilling to run as far as may be like parallel bats of marking, and to be as fine and frequent as possible, both on the individual feathers and on the bird generally. The secondaries white, pencilled across as far as may be like the body.

Golden Pencilled

The same as the Silver Pencilled with the following exception: The ground colour being a rich reddish bay, sometimes with slightly more golden tinge on the hen.

Black Hamburghs

Colour, brilliant, green-black. The beak and legs are often darkers in this than in other varieties of Hamburghs.

Buff Hamburghs

The craze for Buffs has more or less seized upon Hamburgh breeders, though here the disease is not as yet widely spread.

A Buff Hamburgh should be like the other varieties except in colour, which of course should be a rich buff. The ear-lobes should be white and circular; a stain of colour will probably be one of the difficulties in this variety.

RED CAPS

A great egg-laying strain, like inferior Golden Spangled, and with immense meaty combs, the larger the better, provided they are firm in the head, and don't overhang the eye. Ear-lobes, red.

Cocks

Hackles, red striped, blue-black, tinged black at base of neck-hackle. Cock's breast and underparts, black.

Hens

nut-brown, spangled blue-black. These are not much shown in England.

White and cuckoo Hamburghs have been occasionally shown.

POLISH

These birds are so called on the *lucus a non lucendo* principle, because they have nothing to do with Poland. Some think the name a corruption of Polled, a reference to the crest. They have peculiarly formed skulls, having a sort of round upper skull growth in those cases where great crest development is gained, also a strange deficiency of bone formation over the nostrils, accompanying the skull development.

The white-crested blacks have no beard, but largely developed wattles. The spangled have full beard but no visible wattles. The latter are remarkably handsome birds much resembling in colour and markings the spangled Hamburghs. Polish fowls carry the head and neck curiously thrown back. The crest, all important in the breed, should be very large and full; in the cock composed of long thin feathers, as long as possible, like-the hackle feathers in their nature.

In Silver Spangled cocks the crest is white, laced or tipped with black; in Golden Spangled, golden bay, laced with black. The crest should stand well up in front so as not to obstruct the sight. The cocks usually moult a few white feathers in crest when old. The hen's crest should be round as a ball, quite without any split or division. The pullet's crest is black, laced with white or golden bay, but the adult hen has white or bay feathers laced with black. The comb is leaf, red, and *as small as possible;* if absent so much the better. The bodies should be spangled like Hamburghs; the Silver cock has the black spangle at tip of sickles, but the Golden variety has the sickles black edged with golden bay. The hens in each have a black lacing to the tail. The spangling in these breeds approaches more nearly to a lacing than the true spangling of the Hamburgh.

It is said that a weakly chicken will never grow a good full crest, a hint for the selection of the birds for killing.

White-Crested Black Polish

The general characteristics as in the above-mentioned breeds. Crest, pure white, some few black feathers in front, but the fewer the better, forming a narrow band. Rest of plumage, glossy black.

Buff (or Chamois) Laced Polish

Crest and plumage, rich buff, every feather laced or spangled pale buff or white.

There are also white-crested whites, black-crested blacks, white with black lacing, blues, cuckoos, and no doubt other shades. Black-crested whites existed once, and may do so still; if so, are very rare.

Polish won't stand a damp soil, nor does wet improve their crests. They are good layers, but rather delicate.

Langshans

This valuable breed was originally brought over from China, and its name is said to be derived from the place of its origin, *lang*, two, and *shan*, hills. It is said to be "joss," or sacred, among the Chinese, and it must have occurred to some that this cult was nearly taking root among a few in modern England. It has been said that the Chinese regard the Langshan as akin to the wild turkey, but to modern eyes there can be no nearness of kinship except such as may be discussed on such a question as "The origin of species." Except, perhaps, the great controversy which arose on the origin of the Brahma in the Fifties and Sixties, probably no breed has been honoured with so much space in the poultry papers, and so much heated discussion, as the Langshan. Whether the first importation into England was really that to Major Croad in 1872, I am not sure; if so they rapidly got into other hands, as I saw in the hands of breeders of Black Cochins about 1874 or 1875 several hens which they used for crossing, but which were no doubt what is now called the Langshan. They had the general characteristics of the breed, even including the black pupil of the eye and the lightly feathered legs, and I fancy they must have been imported by some one at an earlier date, and called Black Cochins.

The main difference of opinion among breeders of the day seems to be as to length of leg. The determination to have no sign of a Cochin cross remaining–a cross which no doubt existed in some yards–led to the selection of the leggiest and least Cochin-like birds. It is even asserted that in a few yards a Game cross has been used to increase the unlikeness to Cochins, but this, if true in some cases, is no doubt very rare. Still, extreme length of leg cannot in itself be a gain, either for table purposes or as a point of beauty, and no doubt the eventual form preferred will be a bird of moderate length of leg, though far from a short-legged Cochin-looking bird.

The comb should be single and upright. A rather large comb is admired in the cock, but if the combs be grown too large they will have a tendency to fall over in the case of hens, and this is a fatal fault in the show pen. Therefore, though in this breed it is not supposed to be necessary to mate up different pens for producing cockerels and pullets, there is no doubt some slight tendency in that direction, as a fairly large comb is liked in the cock, but it is not always advisable to breed for such large combs for the pullets, lest they should go over.

That Langshans are a most useful bird, either for table or egg producing, is certain. The eggs are rich, large, and more or less brown in shade of shell. The skin should be white and thin. Size, large; an upright and alert carriage. Eyes, dark, with a black pupil. Legs, dark grey (lighter after first year). Scales, in front of legs down to toes, black, with skin between scales and toes showing a pink tinge; under foot, white. Toe nails, white. Plumage, intense black, close and hard, showing a brilliant *green* gloss; a purple gloss being nearly or quite a disqualification. The head should be fine and highly bred. Comb, single, upright, evenly serrated, and free from side sprigs. The back should be fairly long, with close saddle and little or no cushion, and the breast long, full and deep, covered with abundance of white meat. Slightly feathered on leg and outer toes, not *on middle toe*. Tail, large and full, the cocks with distinctly projecting sickle feathers, Yellow tinge on shin, feet, base of beak or around eye is a fatal fault; also vulture hocks, legs quite free from feather, coloured-feathers (not black), comb other than single, too much fluff on thighs, and very short shanks. Occasionally red feathers in hackle will occur in the purest blood, but breeding carefully is eliminating this.

White, blue, and even buff Langshans are claimed as Langshans by some, but are not recognised by most of the best authorities. Still, classes for the two former colours are given at Birmingham.

I have here much pleasure in adding a few notes on the breed which, at my request, were written for this work by one of the most distinguished of Langshan breeders, and will do doubt be of much value to those who are interested in this breed.

"The controversy that has raged about the origin of this breed is a matter of history, and need not be revived. It is less important for us to speculate about the shape of Langshans when they stepped out of Noah's Ark than to know the points for which the jüdges look when the birds are penned for show. And it is less and less necessary for us to hark back to the controversies of the past, as the proper type is rapidly becoming recognised by all, and there are scarcely any exhibitors of importance who do not breed for what has been sarcastically called the 'modern' type. This approaching unanimity will be completed when the Poultry Club's standard of excellence is made public. It has been compiled by a sub-committee embracing all shades of opinion, and the existing standards of the clubs are likely to be altered into line with the Poultry Club standard.

"The Langshan, as imported from China (its home is on the Yangtse Kiang, 150 miles up from Shanghai), is a black bird with green sheen, and excellent for the table, although there are some fastidious people who consider it too large for this purpose. Certainly it may claim to be the best table bird amongst Asiatic large breeds. It has white thin skin, different from all other Asiatics.

"For mating it is not generally considered necessary to have special pens for cockerel and pullet breeding. A few breeders are successful in producing winners of both sexes, but it is noticeable that many run on one or the other and not on both. It is essential to success that the cock for breeding should be as perfect as possible in *brilliancy of colour*, every feather glossed with metallic green (without purple), with clean hocks and thighs, and what is commonly called fluff, lying close as if well groomed. Of course all the points should approach as nearly as possible to the standard of excellence. Some faults in the-cock could doubtless be remedied by mating with hen good where the cock fails, but a cock different in feather colour should not be used to breed prize stock. Another point in both sexes should be carefully watched, *viz.*, the colour of the eye. The eye should be brown-black, and light eyes (which have been too common) are to be avoided. A cockerel with light eyes should never be put into the breeding-pen. It is strange that the eyes of both sexes have a tendency to get lighter with age. Preference should be given to a cock, or cockerel, with rather a small comb, erect, evenly serrated, and not too high in front. The hens, combs should be the same, but of course smaller. The greatest possible fault in comb of either sex is a twist out of the straight.

"In mating, hens of large size should if possible be selected, and cocks of medium size, but both should be well up off the ground, otherwise the progeny will be likely too short in leg for the present fashion. This length of leg is important, in order to keep the Langshan well away from the black Orpington, which has been largely made from the Langshan, and still resembles the parent except in the length of legs, some details of shape, and shanks featherless. The introduction of black Orpingtons into the show pen is probably responsible for the increased height of Langshans. Whether this is so is not important, but it does seem to be important the two breeds should be kept as distinct as possible in type.

"The Langshan is the only breed with feathered legs that has the middle toe free of feathers. The shanks and *outer* toes should have an even fringe of feathers, and this is a point to be carefully considered in mating to breed prize stock."

French Breeds

The French fowls, as eaten in Paris and elsewhere in France, are undoubtedly, if not better than, at least equal to the table poultry of any country. But most of them seem better suited to the dry and sunny climate of France than to our own less-favoured land. They are, however, excellent table fowl, and Houdans are also excellent layers, and very hardy, and Houdan cocks exceptionally prolific.

Houdans

Face, comb, wattles, eye, red; ear-lobe, white or pinky-white; legs., pinky-white, mottled, blue or black.

Cock

Crest, large, full, compact, inclined backwards to show the comb, which is of the type called leaf, rather resembling a butterfly with spread wings. The crest, evenly distributed black and white, generally rather lighter than body colour. (This applies to both sexes.) Beak, horn, short and stout. Eye, red, bold. Wattles, shortish, rounded. Muffles (or whiskers) and beard, large; full, and compact, hiding most of the face. Breast, deep, full, and long. Back, broad and moderately long. Wings, carried well up. Tail, full and arched. Sickles, prominent. Thighs and legs, stout, short, and wide apart. Toes, five, the fifth curving upwards; and starting clear from the fourth toe. The toes should be straight and well spread.

Hen

Crest, very large and globular, compact, upright, and even. Comb, leaf-shaped, and very small. Wattles, very small and round. Neck, medium length and full. Tail, medium, carried well away from body. In other respects corresponding to the points of the cock. Plumage for both sexes, glossy black and white, mottled, and evenly distributed. Young birds are usually darker than old ones. Red or straw feathers, outside spurs, and other than five toes not admissible.

Creve-Coeurs (Split-Hearts)

A name derived from the shape of the combs. They are not so much shown now as formerly. Good layers and table-fowls, but not so suited to our climate as some breeds. They are, apart from colour and comb, somewhat like large Houdans in character, but with red ear-lobes and *four* toes. Back, lengthy, broad and flat; wings, rather large, and moderately tucked up.

Cock

Comb, even, two-horned, V-shaped, not curved, free from branches, moderate in size, and standing well up against the crest. Tail, very full, carried moderately high, with large sickles. Thighs, short and well covered; shanks, short; toes, straight, and long. Carriage, bold. Head, erect. Deaf-ears (in both sexes), small and well covered with muffling.

Hen

Head, rather large. Comb, as in cock; but very small and weak. Wattles, moderate and rounded. Body, square and large. Tail, broad and full. Plumage (in both sexes), glossy green-black. Some adults moult a few white feathers in crest, but these are a defect. Coloured feathers are not admissible.

La Fleche

These are larger, more Spanish-looking fowls, high on leg. Many of the first specimens imported had a slight crest, which is declared against in Mr. Lewis Wright's book, and also by the *American Standard;* but the old *Poultry Club Standard* says: "Crest, very small, consisting merely of a few feathers." A splendid table-fowl, but rather delicate, and not coming quickly to maturity. The head and comb are peculiar. Head, long, thin, flat at the side; comb, like two small round horns, without branches, medium in size, V-shaped. Wattles, long and pendulous, remarkably large as compared with comb and deaf-ear, which is rather small, white, round and flat. The cock's tail is long, not too full; slanting backwards; the hen's; small, and rather erect. Face, wattles, comb, and eyes, red. Plumage, black, with green reflections. Legs, dark slate or black.

Other French breeds are:

Bredas

Various in colour, slightly crested, Polish in character, with large wattles and small deaf-ears. They are not much cared for in England.

La Bresse

Apparently more or les of a Dorking with four toes.

Le Mans

A small-crested Crêve.

Courtes Pattes

Curious single-combed variety, abnormally short on leg, so that the fluff nearly touches the ground, and the bird almost waddles like a duck. The comb and wattles are large. Tail, large, with long sickles. As is usually the case, short neck goes with the short legs. Face and eyes, red. Plumage and legs, black. White ear-lobe. Leg-feather is a disqualification.

Faverolles

This curious muffled breed is as yet but little known in England, and at present it does not seem likely to become very popular. The birds are of a medium size, ranging from about six to nine pounds in weight according to sex and age, and though they have muffling and beards have also five toes. They are practically of two colours, salmons and ermines; though if the present rage for manufacturing buff varieties lasts, which it is to be hoped will not be the case, we shall no doubt sooner or later be afflicted with buff Faverolles!

They should be thick-set massive birds, with shortish heads and single upright combs. Mr. Marx is of opinion that the cock's comb should have only seven spikes, and the hen's not more than five. The ear-lobes are small and round, but hidden by the muffling, which should be soft and very full. The beard is, thick, but of moderate length. The legs should be slightly feathered, as also should the outer toe; the tail only moderate in length.

The combs, face and wattles, red. Beak, horn colour. Eyes, orange. Shin, legs, and feet, and *ear-lobes*, all white.

In *salmons* the *cock* has hackles of straw colour, striped, if possible, with a *brown* shade (a *black* shade is a fault). Back and shoulders, straw and maroon or red. Wing-bar, bronze or bronze-green. Breast and underparts, bronze-black. If mottled with red it is a good point, but not with white. Tail, bronze.

The salmon hen, a dark cream colour, as to head and underparts, running to a dark cinnamon on upper feathers; the neck-hackles-having abroad cinnamon stripe on the cream ground. The hen should be free from black feathers.

The ermines are white all over, except that the, neck-hackles should be striped with black, and the tail should be black edged with white, and are probably too much like bad Light Brahmas to become popular. There will no doubt be many to urge that Faverolles should have salmon or bronze breasts, wings., and shoulders. That day is not yet, and to produce the result some years of careful breeding will be required, while the result may, if attained, be disappointing to the eye.

The principal points to be desired in Faverolles are thus summed up by one of the best authorities.

Size, fine bone, white skin and legs, moderately feathered, single combs, well muffled faces, five toes, and no black in the salmon hens. The eggs in some cases run rather small; they should be large. They are hardy, good layers, good eating, but at present somewhat mongrels.

SPANISH AND MINORCAS

WHITE-FACED BLACK SPANISH

An old and once a very popular breed, now rather eclipsed in popularity by its kindred family the Minorca, which is hardier, bears confinement better, and does not require the trouble taken as to face, which is the chief drawback to an amateur desirous of keeping the "white-faces." There is no doubt that those breeds which can be shown naturally have a great advantage to the majority over those which have to be dubbed, or trimmed, or powdered, etc.

Head, large and deep, especially in the cheeks. Beak, dark horn colour, rather long, not thin. Comb and most of wattles, red, but inside of upper part of wattles and across throat, white. Comb, very large: in cock, *upright*, well serrated: in hen, falling over, almost hiding one side of face. Face and ear-lobes, pure white, like white kid, bare of feathers (with help of tweezers); ear-lobes, very long, pendulous, free from folds, rounded at base. White face, arched far back over eye. True ear, like a black spot in the deaf-ear. Face, as free from folds as possible, and not obstructing eyesight. Neck, long, carried well back. Hackle, flowing. Body, light and active, broad at shoulders, tapering towards tail. Back, rather round and slanting downwards. Wings, long; but closely carried. Breast, round and very prominent; the colour, glossy green-black. The cock's hackles sometimes purplish. Legs, long, slender, clean at hocks, bluish black. Toes, long and thin. Tail, large; sickles, arched, carried high, but not "squirrel." Medium size, with a thin appearance, and carriage upright and strutting, with much movement of head.

These Spanish breeds, with their kindred the Leghorns, no doubt came from either Spain itself or the shores of the Mediterranean, so may be looked on as the contribution of the Latin races towards our poultry classes. The Spanish are generally not hardy, though Mr. Teebay found them hardiest when quite hardened by treating them almost as wild birds, and letting them roost in trees as soon as they can get there, leaving them still to their tree roosts even in midwinter! The size and quality of the white face is an all-important point: the large hairs have to be drawn from it with tweezers, and great care has to be taken to avoid ulceration of the face. The *Black Minorca* is a red-faced Black Spanish.

A BREEDING PEN OF MINORCAS, CONTAINING MANY WINNERS
(Property of Tennyson Fawkes, Esq.)

White Spanish and White Minorcas

Also exist, but the latter are more popular, and the red faces and lobes contrast better with the white plumage.

White Spanish

As in Black Spanish, except that the plumage is pure white.

Spanish are good layers of exceptionally fine eggs, pure white in colour.

Black Minorcas

Much the same as Black Spanish, but face and wattles are red, and ear-lobes pure white, almond-shaped, and medium in size. Wings and legs, medium. Hen's tail not carried too high, and her appearance rather stouter than that of the cock.

Black Minorcas are exceptionally good layers, and few breeds will produce more eggs than they in a year, whether confined in a small run or allowed full liberty. Probably more persons who have only a wired-in run in a back garden succeed with Minorcas than with any other breed.

White Minorcas

Like Black Minorcas, except for white beak, white plumage, and white legs.

Andalusians

A beautiful branch of the Spanish family, but very difficult to breed true. Excellent layers of large white eggs, practically non-sitters, and remarkably taking in colour. The persistent occurrence of black chickens in large proportion is the main difficulty for breeders, and is the result of a Spanish cross early in the Fifties. They were imported from Andalusia in 1851, and many of the original hens had then erect combs.

General Characteristics

Cock

Head, large and deep, tapering. Beak, rather long, not thin. Comb, medium size, single, straight, upright, deeply serrated. Wattles, very long, fine in texture. Face, fine in texture, free from feathers or white stain; very little face over eye. Deaf-ears, medium, close fitting, free from wrinkles. Neck, long, carried well back. Hackle, full. Body, active in appearance, broad at shoulders, tapering towards tail. Back, rather round, and slanting downwards to tail. Wings, long, carried close, and well covered by saddle-hackle. Breast, rounded, and carried well forward. Legs, rather long, free from feathers. Toes, thin. Tail, sickles well arched, carried high, but not so as to be "squirrel-tailed." Size, large. Carriage, very up-right and strutting.

Hen

General characteristics similar, allowing for difference of sex; but the comb falls over.

Colour (Both Sexes)

Beak, dark horn. Eye, bright orange or red, but must match in the show pen. Comb, face, and wattles, bright red. Deaf-ears, pure opaque white. Legs, dark leaden blue. Breast, deep blue. General plumage, deep slate blue, best with a lacing of darker colour, or even of black, on each feather. The neck and saddle-hackles and sickles of the cock, glossy black or deep purple.

The fact that across with the Spanish was resorted to accounts for the difficulty in breeding the blue with the black edge marked enough, without the greater number of chickens, even from the best and best bred blues, coming black. I would suggest that attention be paid to the under-fluff, and that birds both very dark in under-fluff should not be mated together. Though this is only a suggestion, it might be worth trying, and the results worth noting.

The following notes are sent me at my request by one of the best known and most successful breeders of the variety:

"Andalusians are a sport from the Minorca, probably the result of a cross between black and white birds. They are still called Blue Minorcas in the West of England. They are hardy, stand confinement well, and as a rule feather quickly, but occasionally one of a brood will be almost bare of feathers and will somewhat. resemble Dumpies. These birds are well worth watching, as they often come with clearer ground colour and better lacing than the majority of the chickens."

"They are prolific layers of large eggs, but for table are only of moderate value."

"When first exhibited they were of a uniform rather pale slate colour, but now their chief beauty is the sharp black lacing on a blue ground. To get this in pullets, which is most difficult, it is necessary to mate up heavily laced birds on both sides, that is, the cock and the hens should all be heavily laced, and have the same tone of ground colour. In fact for breeding show pullets it is quite useless to mate up birds that are not heavily laced on the wings, and the hens should also have the tail feathers well laced."

"For breeding cockerels the case is different, and many of the best cockerels are bred from dark cockerels mated with lightish hens."

ORPINGTONS

A modern composite breed, for which Mr. Cook, of Orpington, was primarily responsible. They are, no doubt, a fine laying breed, and of good table qualities. Originally black, at first single, then also rose-combed, they have now developed a buff variety, which, like other buff breeds, seems likely to hit the taste of the day. They are layers of brown eggs, and were originally much like a clean-legged, short-legged Langshan. In fact, pure Langshans have been shown successfully in Orpington classes in their early days.

They still occasionally, revert, throwing a little feather on leg, or coloured or white feathers, but careful breeding will get rid of these difficulties. The original compound was Minorca, Plymouth Rock and Langshan. They have white skin and flesh, red face and ear-lobes, wattles well developed, comb either single (not so large as a Minorca's) or rose. The body colour, rich green-black. Legs, short and black. Back, short; saddle, rising slightly; fairly large 'tail; the sickles of the cock well curved. Eye, black, with dark brown iris. Feet, black; four toes; nails and soles of feet, white. Hen has a slight cushion; tail, somewhat rising.

BUFF ORPINGTONS

These are in type similar to their black kindred, but with single comb and white legs.

With regard to these birds a first-class controversy has raged, as to whether they have any fair claim to be called a distinct breed at all, or whether they are not practically the same as what have for

BLACK ORPINGTON COCKEREL
Winner of 1st and Cup, Crystal Palace, 1901
(Property of G. Duckworth, Esq.)

some years been known as "Lincolnshire Buffs," a manufactured race from the Cochin.

That there is great similarity, I think few unbiassed persons can deny; but as the controversy is still undecided, the reader can consider the matter for himself, and if so minded rush into the fray and give and receive his share of the hard knocks going. The original claim made for Buff Orpingtons was that they were manufactured from a triangular mating of Dorkings, Hamburghs, and Buff Cochins.

WHITE ORPINGTONS

These are either rose-combed or single-combed, but resemble the blacks in all points exccpt that they have pure white plumage and. *white legs.*

Leghorns

A valuable American-made variety, containing, no doubt, much of the blood of the Spanish and Minorca breeds, all natives of the shores of the Mediterranean or of Spain. In fact the White Leghorn is the White Spanish but with yellow legs. (The Americans, unlike the English, prefer the yellow leg for market purposes). Leghorns are excellent layers of white eggs of good size, are easily kept in confinement, and are hardy, useful birds, though not of the first quality for table according to some opinions. The Americans at first preferred the tail of a Leghorn cock to be carried "squirrel fashion," but this is nowadays quite decided against, at any rate in England, while the American standard says "carried upright." The Brown Leghorns are larger, as a rule, than the whites, and were no doubt originally brought from Leghorn. It is important in breeding to keep in view the white ear-lobe and yellow leg. They are, or should be, non-sitters. The later varieties, Duckwing and Pile, were no doubt suggested by the resemblance in colour of the Brown Leghorn cock to the brown-red Game cock, and the idea of crossing, as in Game, to get the other colours, occurred to some ingenious mind. Buffs are a more modern growth, and this fancy must have been a little gold mine to breeders of inferior *Buff Cochins*. It is not very unusual to find a somewhat Cochin head on an otherwise good Buff Leghorn. The question of colour-feeding, so well known in the canary world, is now coming up for discussion, and it is no doubt among the Buff-breeds that the matter is principally of importance. The American standard also includes Black and Dominique Leghorns.

General Characteristics

Head and neck, of the Spanish type. Comb, face, etc., red. Comb, single, straight, erect, large, deeply and evenly serrated, free from side sprigs, and extending well over back of head, fine in texture. Wattles, fine, long, and thin. Face, fine in texture and free from folds. Deaf-ears, rather large and pendant, thin, close-fitting, and free from folds, white resembling white kid, but a creamy tinge is allowable in buffs. Neck, long, curved upright. Hackle, full and flowing. Body, active looking, wide at shoulders, and narrowing towards tail. Back, rather round, sloping towards tail. Wings, large, well clipped up. Breast, full, round, and prominent. Legs and thighs, rather long; shanks, slender, quite free from feathers; legs brilliant yellow. Tail, large, with full sickles, carried high but not so as to be squirrel-tailed. Size, medium, but the larger the better. Carriage, sprightly. Shape, of the Spanish, not Dorking type. Eyes, red, but in buffs orange are also allowable.

White Leghorns

Beak, yellow. Plumage, pure white, of a blue tinge, not yellow.

Brown Leghorns

Beak and toe-nails, yellow or horn colour.

Cock

Head and hackle, rich orange-red, black stripings; crimson tinge in front of hackle. Back, shoulder, and wing-bow, deep crimson-red or maroon.. Wing-bar, blue-black, green reflections; primaries, brown. Secondaries, bay outside, black inside. Saddle, rich orange-red; sometimes a little black striping. Breast, thighs, and underparts, glossy black (no brown splashes). Tail, black with green reflections, no white in tail. Tail coverts, black with brown edgings.

Hen

Hackle, rich golden-yellow, broad black stripes. Breast, etc., salmon-red; round head and wattle, maroon. The colour shades into ashy grey at the thighs. Back, wings, and shoulders, rich brown, very closely pencilled with black. The wings to be free from any red tinge. Tail, black, with brown pencilling.

Pile Leghorns

Beak and toe-nails, yellow or horn colour.

Cock

Head and hackle, chestnut-red. Back, deep chestnut, almost claret under neck-hackle. Wing-bow, brick-red. Wing-bar, white, faihtly edged chestnut; secondaries, chestnut outside, creamy inside. Saddle, chestnut. Breast, creamy, white under throat, slightly laced with chestnut. Tail, white or creamy white, as free as may be from black splashes.

Hen

Hackle, pale chestnut, centre of each feather slightly white. Breast, rich chestnut, under parts raler. Back, creamy white, faintly laced chestnut. Tail, white.

Golden Duckwing Leghorns

Beak and toe-nails, yellow or horn.

Cock

Straw-colour, deeper towards front. Outside feathers as free from black stripes as possible. Back, saddle, and saddle hackle, deep gold; hackle, rather paler. Shoulder, solid bright gold. Wing-bar, bluish violet; primaries, black, outer web edged white; secondaries, white outside, inner web and lower butts black. Breast, under parts, thighs, and tail, black with green reflections. Grey fluff at root of tail.

Hen

Head and hackle, silvery white, sharply striped black or dark grey. Back, saddle, and wings, clear steel-grey, finely pencilled with darker grey or black, showing no shade of red or brown. Breast and under parts, deep salmon-red, darker at throat, shading to ashy grey or fawn on the under parts. Tail, grey, rather darker than the body. Inside feathers, dull black or dark grey.

Silver Duckwings

Cock

Head and hackle, silvery white, outside feathers, as free from black stripes as possible. Back, saddle, saddle-hackle, and shoulders, solid silvery white. Wing-bar, bluish violet; primaries, black, outer web edged white; secondaries, white outside, inner web and lower butts black. Breast, thighs, under parts, and tail, black, with green reflections. Grey fluff at root of tail.

Hen

Head and hackle, silvery white, sharply striped black or dark grey. Back, saddle, and wings, clear silvery grey, finely pencilled with darker grey or black, showing no shade of red or brown. Breast and under parts, light salmon-red, darker at the throat, shading off to ashy grey or lawn on the under parts. Tail, grey, rather darker than the body. Inside feathers, dull black or dark grey.

Buff Leghorns

Beak, bright yellow (or horn). Eye, red *or orange*. Ear-lobes, white *or creamy white*. Legs, orange or yellow, *free from any suspicion of green*. Toe-nails, *white*.

Note that in some important points the Buff standard differs from that of the other varieties, apart from difference of plumage.

Cock

Plumage, any uniform shade of buff, from rich lemon to orange, the neck, saddle-hackles, and wing-bow being of greater lustre. Tail, rich clear buff, chestnut, or bronzy chestnut; the former preferable, harmonising with the body colour.

Hen

Plumage, including tail, any uniform shade of buff, from rich lemon to orange. Hackle not conspicuously darker than breast.

The main faults to guard against in Leghorns are: twisted or falling combs in cocks; erect combs in hens: red ear-lobe; miscoloured legs; coloured feathers in Whites; white feathers in Browns; black, white, or black and white feathers in Buffs; also a Cochin's head instead of the Spanish type in Buffs, and feather on the legs in the same variety. The use of the Buff Cochin as a cross for colour is no doubt the cause of these latter difficulties.

Note: The above represents the views of the Leghorn, Plymouth Rock, and Andalusian Club, which may be considered as authoritative, until the New Standard is issued by the Poultry Club. A few other colours are also bred, such as blacks, black and red mottles, and cuckoos, which seem to be practically Anconas.

Plymouth Rocks

The Barred variety of Plymouth Rocks was original, manufactured in America and imported into this country.

Now, with the craze for new strains, and especially of white and buff colours, it is not surprising to find that White and Buff Plymouth Rocks are recognised. The first Plymouth Rock strain, that of Dr. Bennett, in America, practically disappeared, and the modern bird is a larger, single-combed Dominique, but with a very small tail of the Cochin type. No doubt they were the produce of Dominiques and Asiatic breeds, including, probably, both Cochins and Brahmas. They are excellent fowls, both for laying and table purposes, and, as a rule, lay a good-sized brown egg.

In breeding cuckoo-marked fowls the breeder will be at times troubled with pure white, black, or red feathers. The latter are far worse than the other two, as they are quite at variance with the tone of the plumage. I can find no clear statement in any English Standard as to the *colour* of the eye or a Plymouth Rock. The American Standard says "bright *bay*." I should think that a red or reddish eye would be preferred, though a pearl eye should not be a disqualification. The Americans have also a strain of Barred Rocks with pea combs. The origin of this may be, no doubt, attributed to the Brahma.

General Characteristics

Cock

Size, large. Comb, single, medium, straight, with well-defined serrations. Beak, thick and strong. Head, medium, and carried well up. Eye, large and bright. Wattles, rather long and large, well-rounded. Ear-lobe, well-developed and pendant. Body, large, square, and compact, with short, broad back. Wings, medium, carried well up; butts covered by breast and saddle-feathers. Legs, rather long and bony, standing wide, and free from feathers. Feet, stout and strong; four-toed. Tail, very small, like a Cochin's. Thighs, stout, carriage upright.

Hen

Comb, beak, head, eye, wattles, etc., same as in cock. Back, broad; the neck-hackle flowing weli over shoulders. Tail, small, upright, rather pointed. Thighs, stout and well-fluffed.

Barred Plymouth Rocks

In both sexes the comb, wattles, and deaf-ears are brilliant red. The beak and legs, bright yellow. The plumage, cuckoo (see *Cuckoo Cochins, Cuckoo Dorkings*, etc.), namely, pure light grey and black, the shafts of the feathers corresponding with the bands. No red, black, white, or yellow feathers.

White Plymouth Rocks

In other respects like the Barred variety, but the plumage pure blue-white, free from yellow tinge.

Buff Plymouth Rocks

In other respects like the two foregoing, but the eyes have a bright bay iris.

Cock

The plumage, any shade of buff, uniform throughout, from lemon to rich orange, the hackle, saddle, and wing-bow showing greater lustre. Tail, rich buff, chestnut, or bronzy chestnut (the former preferred) to harmonise with the body-colour. *Under-colour* also to harmonise with surface-colour. This is a point to note.

Hen

Plumage, any uniform shade of buff as above, the hackle not conspicuously darker than the breast. Tail, same shade of buff as the body.

Fatal defects include feathered legs, legs not yellow, rose combs. Red, black, or (except in whites) white feathers. White ear-lobe.

PLYMOUTH ROCK COCKEREL
Winner of Silver Cup, value 5 guineas, Crystal Palace, 1901
(Property of W. Slater, Esq.)

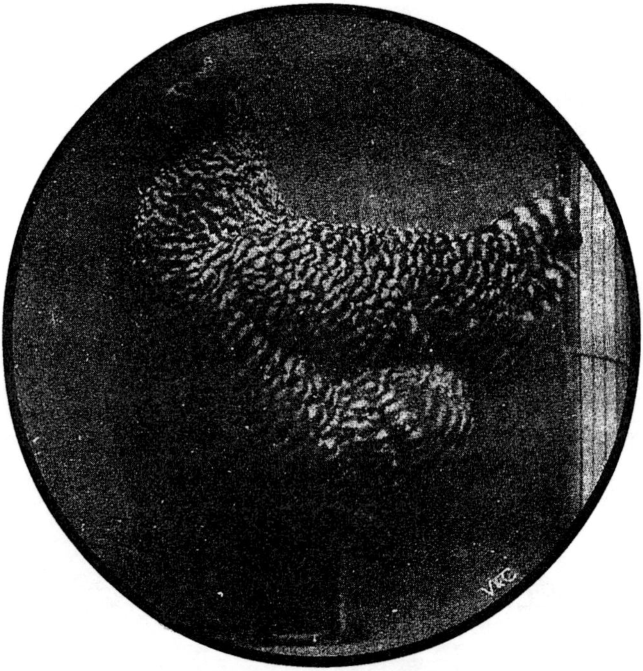

PLYMOUTH ROCK HEN
Winner of the Diamond Jubilee Silver Cup, value 5 guineas, Crystal Palace, 1901
(Property of G. Jackson, Esq.)

Coloured feathers in Whites; also in Buffs. In cocks, spotted hackle (neck or saddle), black or white in wing. White in tail. In Buff hens, black spots or stripes in hackle. Feathers other than buff in wing; white in tail.

Dominiques

One of the oldest American breeds, of the cuckoo style of marking (see *Cuckoo Dorking*), resembling a rose-combed Cuckoo Dorking, but with four toes and a yellow leg. The comb, a rose-comb of the Hamburgh type; ear lobes red, legs yellow. They are excellent layers and table fowl, and hardy.

WYANDOTTES

Another of the comparatively new breeds, made in America, excellent layers and fine table fowl. Several varieties are now bred, and they stand confinement well.

SILVER WYANDOTTES

Cock

Head, short and broad. Comb, rose, firm on head, red; top, oval, and surface covered with small corrugations; small spike at back. The whole comb and spike slightly curved to follow the line of the skull. Wattles, red, medium, fine. Ear-lobes, red (white in ear-lobe a bad defect), well developed, and smooth. Beak, horn, shading to yellow. Neck, short, arched; hackle, flowing, silvery white at the point, with narrow blacks stripe, clear edges of white. Back, short and broad, white. Saddle, broad, rising with concave sweep to tail. Saddle-hackle, white, with narrow black stripe as in neck-hackle. The hackles sometimes show a tapering centre of white in the Black. Breast, black feathers, with large pointed white centres, forming a sharp black edge to the breast feathers. Body, short and deep; web, of feather, dark slate, sometimes slightly frosted with white, but best without. Under-colour, dark slate.

The English view is rather of a white feather, running dark slate downwards to skin, with edge of green-black running well round the feather, free from streak at point.

Secondary wing feathers: visible half, white, running to upper edge, black. Wing coverts: upper web, black, lower web, white, with narrow black stripe along the edge, widening towards the tip, forming a double-spangled wing bar. Tail, black, well developed, wellspread at base. Sickles, beetle green, black, well curved, projecting slightly beyond the true tail. Tail coverts, black, the lesser coverts edged white. Thighs, stout and short, covered with black (not so good), or grey powdered on black feathers. Legs, yellow, short and stout, free from feathers or stubs on shanks or toes.

Hen

Generally of similar character to the cock. The neck-hackle, white, narrowly striped black, may show white shaft of feather. Back, short and broad, slightly cushioned. Feathers of back, black, with medium white centres, the black to have no white edging. The white

CHAMPION SILVER WYANDOTTE COCK
("The Old Genteleman")
Claimed to have been the finest Wyandotte ever seen.
(Property of W.A. & R.F. Spencer)

to be free from pencilling, and the black lacing sharply defined. Breast, white feathers, sharply edged black. No pencilling in the white parts, and the black edging not to be again laced with white. Under parts of body: black web, of feather, with narrow white centre running into black, or black frosted with white near thighs. Fluff, full and rounded; dark slate, powdered with grey.

Secondaries: outer half, white, inner, black. Wings and shoulder coverts, white, heavily laced black, the black lacing widening over the wing-bow. Tail and greater coverts, black. Lesser coverts, black, or black with narrow white centres. Under-colour, dark slate.

Golden Wyandottes

The same characteristics as the silvering substituting for white the following:

Cock

Hackles, golden bay, also on body and wings. Wing-bows, deep reddish bay. Shoulder coverts, rich deep red. Thighs, powdered with yellow. Under-colour of breast, slate, slightly tinged yellow. Tail coverts, edged golden bay.

Hen

Corresponds to the silver hen, except for the golden ground colour, etc., to take the place of the white.

White Wyandottes

Plumage, pure white throughout. Legs bright yellow. Beak, clear bright yellow.

Black Wyandottes

Plumage, pure black, the hackles, back, and coverts showing greenish. gloss. Legs, black, or black shading into willow or yellow. The bottom of the feet, yellow. Beak, black, shaded yellow.

Buff Wyandottes

Beak, legs, and toes, yellow. Plumage, rich, deep, clear buff, uniform throughout except tail, a rich, deep buff or coppery bronze. Under-colour as surface, but may be rather lighter, and the colour should extend to the skin.

Partridge Wyandottes

These birds should in other respects resemble the ordinary Golden Wyandottes, but having the colours of the Partridge Cochin. These may shortly be described as having reddish orange hackles, both neck and saddle, densely striped with black down centre of feather: a white shaft down centre of feather is most objectionable. The neck hackle of the hen may have pencilling on it similar to that on her body.

The cock is of a rich dark red on the back and shoulders showing bar, black; secondaries, outer edge, rich bay, inner edge, black. Breast, under parts, and tail, rich black.

Hens should be of an even brownish shade, hackle more golden but as near to the shade of the body as may be. The whole of the plumage to be pencilled ove - with fine markings, following more or less the outline of the feather, and resembling the pencilling of a well-marked Brahma.

Buff-laced Wyandottes

These are curious-looking and rather pretty birds: if the breeders still improve the accuracy of the lacing, they should become very pretty. The beak and legs are yellow.

Cock

Neck-hackle, buff, with white stripe. Breast and thighs, rich buff, with sharp white lacing to each feather, and white under-colour. Saddle-hackle, rich buff, with white striping down centre of feather. Shoulder and wing, solid buff, like saddle in tone. Wing-bars, buff-laced white. Secondaries, tail and Coverts, all white.

Hen

Generally resembles cock, being laced with white (like the breast of the cock) all over the body. The tail, white. A black tail is a bad fault, also blue shade of lacing.

The received view is as above, but the white tail as a great blot in the harmony of the bird, and if a buff-laced tail were the accepted standard it would both improve the appearance of the bird and probably enable the strain to come more easily true to colour.

Frizzled Fowls

These may be of any variety or colour, provided that instead of the feathers lying flat as in ordinary breeds of fowls, every feather is turned backwards, as if curled back. They have been shown of several colours, black, white, brown, greyish buff among others, and Frizzled Bantams have also been manufactured in one or two instances. They are not bad fowls, either for eggs or table, but in my view more curious than beautiful.

Sultans

A curious and pretty breed of white fowls, brought from Constantinople, rather resembling White Polish, but with feathered legs and feet, and vulture hocks. Good layers and non-sitters. They have large crests and muffs, *five toes,* full plumaged hackles and tail. The comb, merely two little points; pendulous wattles (in the cock). They seem to stand half-way between ordinary fowls and bantams. These are scarce, few breeding them, and fresh blood is very hard to get. It is to be hoped they will not be allowed to die out.

Silkies

These curious and beautiful little fowls are of old pedigree, and are remarkable for the colour of the skin and the silky hair-like nature of the plumage. They are of Eastern origin. They are not, perhaps, profitable poultry in the ordinary sense, but make capital sitters for

SILKIE COCKEREL
Winner of 1st Prize, Crystal Palace, 1901
(Property of Mrs. A. Campbell, Esq.)

pheasant or partridge eggs, and are good mothers. The head and beak, small. Face, dark purple. Comb, a queer, lumpy round rose comb, of dark purplish colour. Crest, full and round, not cockatoo shape. Wattles, long, purplish. Ear-lobes, purplish, tinged white. Body, square, covered with silky fluffy feathers. Wings, carried low. Legs, shortish and moderately feathered, but not vulture hocked, bluish black in colour, the outer toes being feathered, and five in number. They are said to be good eating in spire of the purple colour of the skin.

Dumpies

A very short-legged breed, the shanks being about 1½ inches long. They are somewhat like stunted light Scotch greys, sometimes with more of a brown tinge. A hardy and good table fowl.

Rumpless

These are simply fowls which naturally have *no tails,* and are a very old breed, known for some centuries. They have not even the physical formation on which to hang tail feathers. In animals, Manx cats are similarly tailless, and some few breeds of dogs, such as Schipperkes, are likewise afflicted, though not breeding so true as the fowls and cats. Size, rather small. Combs, either single or rose, and great variety of colour, though white or black are preferred. Some of the best have been five-toed, feather-legged and vulture hocked, crested and bearded. Their odd appearance is the main recommendation of this queer variety.

Yokohamas

A curious variety, requiring specially large pens for exhibition, owing to their long tails, which sweep the ground. These or similar birds are also called "Phoenix" fowls. They are of gamey appearance, and of Game colours, and it is stated that the, tails (in Japan) are only moulted once in three years, and that a tail has been measured over twenty feet in length! In England a six-foot tail is considered a good one.

Bantams

The ingenuity of man, especially of the China-man and Japanese, and also of the late Mr. Entwistle, has produced dwarfed specimens of most breeds of fowls, which are valued as pets or curiosities, though not strictly suited for poultry farming for profit. Game, Cochin, Brahma, Polish, Malay, Andalusian, and other breeds are shown, dwarfed from the large breeds, which they resemble except in size, and Frizzles of various sorts are also in existence. In time probably every large breed will be represented in Bantam form. A few breeds require special mention.

Black Rose Comb

Some division of opinion existed at one time as to the correct type for these, but it is now generally admitted that the bird should be in all respects a miniature Black Hamburgh.

BLACK ROSE COMB BANTAM COCKEREL
Winner of the 5 and 10-Guinea Challenge Cups, Crystal Palace, 1901
(Property of Lady Alington, Esq.)

White Rose Combs

These are also shown occasionally. They should have white beaks, white or pinkish legs (in America also yellow are permitted), in other points resembling the blacks.

Booted Bantams

The ordinary colour is white, but blacks have existed, and may do so in small numbers still. The comb should be single, though rose-combs have been shown, and upright in both sexes. Beak and leg, white; the leg as heavily feathered as possible. Ear-lobes, red. Tail, upright. Wings, large and slightly drooping. Plenty of sickles and coverts. Shanks and back, short.

Japanese Bantams

Pompous-looking little white birds, with large black tails laced with white, carried erect. Legs, very short and yellow. Some Japanese of a cuckoo or speckled colour have also been shown.

Cuckoo Bantams

These are really dwarfed Scotch greys, with single comb and white legs. Some Cuckoos with rose combs have also appeared.

Nankin Bantams

Rarely seen as small as other varieties of Bantams.

Cock

Like a very dark Buff Cochin, a sort of dark orange on upper parts, but instead of a Cochin tail a large glossy black tail with a sweeping carriage.

Hens

Buff Cochin in colour, but with dark hackle (sometimes spotted with black), and black tips to tail. The clearer the neck-hackle the better. Combs, single or rose. Legs, dark., of a blue shade, which often become white.

Pekin Bantams

The Cochin, or Pekill Bantam, found in the sacking of the Summer Palace at Pekin in the China war of 1860, have become more popular than Nankins, being miniature Cochins.

Probably there are no more fascinating birds than Pekins in the whole range of Bantams; their tameness and their round fluffy appearance attracting everyone. They are difficult to breed with a sufficiently small tail, and it is openly asserted that the great majority of Pekins are exhibited with but a moderate proportion of their natural tails, the larger and stiffer feathers being removed by violence to give the proper soft and Cochiny appearance. At any rate, it is wise for novices, when buying stock, to examine carefully whether there appears to be any unnatural deficiency in the main tail feathers, and to make sure that feathers have not been removed. That this warning applies to judges also is obvious.

The *Buff* Pekin should be as far as possible simply a miniature Buff Cochin. The smaller and the more thoroughly *Cochin* in character the better, a statement which applies to all Pekin Bantams.

The *Partridge* Pekin resembles the Partridge Cochin, and the show cock should have the glorious colouring described as belonging to the Partridge Cochin cock, while the hen should be laced all over as the Partridge Cochin hen. To obtain both sexes in perfection it is necessary (though some may dispute the assertion) to have different breeding pens: one to breed the show cockerels, whose sisters will be very dimly marked or not at all, and more slate-coloured than the show hen; the other to breed the show pullets, whose brothers will be too much laced or splashed on breast and fluff to be show birds, but will be most valuable in their turn to breed show pullets.

The *White* Pekin should be of the purest blue-white all over. If it should unfortunately be necessary to cross these with any other colour, it must certainly be with Blacks, not with Buffs or Partridge, or a strain of bad white will result, which it will take years to get rid of; and a good green-black should be selected of a strain which it is known has not for many generations been itself crossed with Buffs or Partridges.

Black Pekins are miniature Black Cochins; and no doubt could in some instances be produced by crossing Buffs and Whites, as was the case in the ordinary Cochin. But such crosses, unless absolutely necessary should be avoided, as no care is too great to keep the purity of strain and purity of colour when once obtained. Blacks crossed with Buffs would be likely to throw red or bronze hackles; crossed with Whites, white feathers might be looked for, especially after adult moult.

Cuckoo Pekins are not as yet common, but are very pretty when good. For the characteristics of cuckoo marked birds, see under Cuckoo Dorkings.

Again avoid crosses if possible. If, however, the colour becomes too light, and cannot be regained by selection of the darkest birds for breeding, then a cross may be necessary. This should be between a cuckoo cock and a black hen of black or black and white blood, free from any admixture of buff or partridge strain. The pullets from such across crossed back with a cuckoo cock would probably the desired colour.

SEBRIGHT BANTAMS

These lovely little birds are among the most wonderful products of the breeder's skill, and were manufactured by the late Sir John Sebright. He began the process in about the year 1800, crossing between Bantams and Polish; then accidentally came across a hen-tailed Bantam cock, no doubt with some henny game blood in him. The object was to get tiny Bantams with the beautiful Polish lacing, but with henny hackles and tail in the cocks. Truly a difficult thing to attain, yet it has been perfectly attained, and no doubt Sir John would be astonished if he could see the perfect Silvers and Golds shown nowadays.

The Sebright Bantam is rose-combed. Comb, wattles, and face, red. Ear-lobe, best white, but red admissible. Beak, short, horn colour. Eyes, full, bay. Neck-hackle hen feathered in both sexes, as also must the back and saddle be. Wings, large, carried low. Shanks, slate or blue. Tail (in cock) henny, rather expanded, carried high, quite free from sickles. The upper feathers of the tail to extend as little as possible beyond the rest. Coverts also henny, straight, round at ends, and lying close to the tail. Colour: in Silvers, a pure white ground; in Goldens, a rich golden yellow; in each case *every* feather to be sharply and evenly laced with a narrow black edge. When excited, the head and tail almost touch, and both head and neck have a tremulous motion such as is seen in the fantail pigeon. The eggs were very unfertile, but crosses and an increased number of breeders have tended to improve matters.

GUINEA FOWLS

Guinea fowls have two or three drawbacks to go against their good points: one their tendency to s[illegible], and also to lay away; the cantankerous nature of the cock towards other poultry or game, it having been found too spiteful to pheasants to be allowed to live wild as a game-bird; its delicacy in the earlier stages of its life before the horn is fully grown. The similar delicacy of young turkeys until they have "shot the red" is well known. It is suggested that any one starting Guinea Fowls should obtain eggs and hatch them out, as adult birds would be too wild, and unlikely to stay. The eggs are rather small and pointed, creamy in colour, and excellent eating. As to the meat, opinions differ. Some think very highly of it, others think it rather dry. No doubt the more the birds run wild the better the quality of the flesh, and they should be well hung, like pheasants. Mr. Wright recommends that, except the nest egg, the eggs should be removed daily, as if they are allowed to accumulate, and are then removed, the nest will be deserted. More than one hen will lay in the same nest, and the period of incubation is twenty-six days. Some declare that for success the Guinea Fowls should be in pairs only, but two hens to a cock are usual: even four or more hens have been successfully tried, but the experiment is risky.

The Common Guinea Fowl

This has a horn, red wattles, white under eye, and white ear-lobe, neck-hackle a brownish grey. The body grey, speckled all over with white spots, the grey running to purplish black in places. Legs, dark or slate. Carriage of stern very low; the profile of back tail appearing almost like a segment of a circle. In some cases the colour is reversed, and is greyish white, speckled purplish.

White Guinea Fowls

These are not uncommon, and have not the stigma of delicacy attached to them which is attached to most white breeds, however unjustly. Only in a *White* Guinea Fowl is a white feather permissible.

Pied Guinea Fowls

Produced by crossing of whites with the darker varieties.

Turkeys

The old breeds in England were generally known as (1) the *Norfolk Turkey,* which is black, spotted white on wings; excellent in quality, but not so large as (2) the *Cambridge Turkey,* a bronze-grey bird, larger and leggier than the Norfolk. The fine American bronze turkey is now much liked, and is used with advantage for crossing with the other strains. White turkeys are generally thought more delicate than other colours; but as this is often believed of other breeds of white fowls and animals, and sometimes without good ground, it may possibly be in this case also a sort of superstition.

The general appearance of the turkey is so well known that beyond saying that birds of slate, buff, and black colours exist, it is not necessary to describe them.

They are generally rather delicate as chickens, especially at the stage called "shooting the red," and damp is considered very bad for them at that time; but I have seen strong evidence that in some cases turkeys do well even in damp fields.

They have been judged mostly for weight, though beauty of plumage and condition should not be overlooked. The American turkey *is* mostly a combination of rich bronze, edged black, or black shaded bronze. In all breeds it is important that the breast bone should be *long*.

Period of incubation, 27 to 28 days.

Ducks

For practical purposes in England, the breeds of ducks may be called three in number, or at most five.

The Aylesbury Duck

This is remarkable for its table properties, in which it is unsurpassed. It is large, hardy, and matures early, and thousands annually come to the London market. Colour, pure white, a yellow tinge suggesting a cross with Pekins. The drake has a curl on the tail. Some sort of pond or running stream is generally considered necessary for successful duck breeding, but the experience of several breeders, especially of some breeders on a very large scale in America, goes to show that the necessity is much exaggerated, and that it is perfectly possible to breed ducks in quantity without any water for them to swim in. If an artificial pond is made, take care that the sides are not too steep, or the young ducks may be drowned, exhausted in the effort to get out. The head should be long. Bill long, broad, and in profile running nearly straight from tip of bill to top of head. The bill should be pale flesh (not yellow) in colour, and quite free from discolouration or marks. Eyes deep blue. The body should be long and deep, as much so as possible; and the increased depth obtained of late years has quite altered the appearance and carriage of the Aylesbury, to its manifest advantage.

ROUEN DUCK

A larger and tamed wild duck. Few birds are more handsome than a Rouen drake, and the breed is most profitable as a layer, but later in breeding and maturing than the Aylesbury. The drake should have a greenish yellow bill, with a black "bean" at the tip, long, broad, wider at extremity than at base, rather straight in line with the head, as in the Aylesbury. Head, glossy green. Eye, dark hazel. Neck long and slender, glossy green, with white ring at lower part not quite meeting at back of neck.

Breast, rich claret, the colour extending as far down as possible, and as free as possible from rings, the well-known "chain armour" of breeders. The under parts are a beautiful grey, the colour running right under the tail, where it becomes black. Back, ashy grey and green, rich green towards tail, the curl of the tail greenish black. Wing-bar, bright blue, edged on each side with white. Legs, orange.

The duck has a brownish orange bill, with dark blotch, and a black bean on the tip. Her neck must have no sign of a ring. Head, deep brown, with two lighter stripes on each side, running from the bill, over the eye, and some way down the neck. Breast, brown-pencilled. Back, lighter brown, marked with green-black. Under parts greyish brown, pencilled darker. Wings have the blue ribbon bar edged white.

The drake changes, like the wild duck, to duck plumage in summer.

PEKIN DUCKS

These have rapidly come into favour in the last fifteen years, and are hardy, good layer, and excellent eating. In fact, the finest flavoured duck I ever tasted were some Pekins belonging to the late Dean of Jesus College, Cambridge, Rev. E.H. Morgan, which had been reared on *a very dirty pond.* I hardly like to consider how they attained so fine a flavour.

Pekins are of a uniform pale straw or creamy colour, with deep yellow bills of medium size, and quite free from discolouration or marks.

They have long deep bodies, short wings. Tail carried high, and the drake's curls rather hard and stiff. Shanks short, stout, and reddish orange. The plumage is very downy.

Black Cayuga

Colour, lustrous black throughout, the drake being far more brilliant than the duck, and in parts a rich green. The duck's primaries are sometimes dark brown. The shanks a dusky orange, dark slate, or black, often covered with a network of black lines. The bill is rather long, broad, and dark, a blue-black with a darker blotch preferred. Eye, dark hazel. The body, while long, is round and plump. Wings long. It is difficult to get as brilliant a shade on these as on the East Indian. Some say the neck should be lengthy; the American Standard says medium, and the bill rather short. They reach as great size as the Aylesburys and Rouens, are hardy, good layers, early maturing, and of fine flavour.

Black East Indian

A brilliantly coloured bird, of small size, feet and legs as nearly black as possible. The drake's bill should be of a yellowish tinge, with an apparent thin wash over of green. The duck's bill should be black, or nearly so. They off moult a few white feathers round eye, at root of bill, and on breast; these are defects, and the fewer there are the better. Heads, small, and high-bred looking. They cannot be too small in size for show purposes; are good layers, and carry a great deal of meat in proportion to their small bone.

There are many other varieties of Ducks, such as Indian Runner, Grey-Call Ducks, White-Call Ducks, Muscovys crested white, Whistlers, both the red-billed and the white-faced, and the more especially ornamental sorts, whose colours are more brilliant than one could imagine without seeing them. Such are the quaint Mandarins, Carolinas, Japanese Teal, the falcated Ducks, and Bahamas; but these are not kept merely for practical purposes, but as ornamental water-fowl.

Ducks (unlike fowls, which should be plucked as soon as possible after they are dead, and while still warm) should be plucked *cold*. The story of the boy who, being used to plucking ducks, was set to prepare a hare for the table, and found it almost impossible to *pluck*, may occur to the reader.

Geese

An ancient race, celebrated in history as having by their cackling aroused the Romans, and saved the Capitol from the attacking Gauls. Of domesticated geese there are two principal kinds, the Toulouse, called for some reason after the town in the south of France, and the Embden, so called from the town of Embden, in Westphalia.

The Toulouse

These are a handsome and compact kind, as large as any when tested by weight, though shorter in the leg than the Embden. Bill, rather short, reddish flesh-colour. Upper-colour, dark grey. Breast and under-colour, lighter grey, becoming white on belly, fluff and rump. The white is scarcely visible when the bird is viewed from the front. The back is somewhat convex. Wings, large. Tail, rather short, grey and white, ends tipped white. Shanks, short, reddish orange. The body of a bird in good condition almost touches the ground. Eye, bright hazel or brown.

The Embden

Plumage, pure white throughout. Bill, medium length, stout, dark flesh-colour. Eye, bright blue. Carriage, very tall and erect, with square deep bodies, close to ground in fat birds. Wings, large. Tail, rather short and stiff. Shanks, deep orange.

Three-year-old ganders have weighed 36 lb., and geese 26 lb. or over, but these are exceptional weights. A gander should, when adult, weigh 20 lb. upwards, a goose 18 lb. upwards. Two geese are enough to mate with a gander, and it must be remembered that geese rarely lay till after a year old. Geese sometimes lay two sittings in one season. Period of incubation, 30 days.

From geese, kept hot and overfed till they acquire enormous livers, is obtained the material known to epicures as *pâté de foie gras*, pleasant enough to the palate, but unpleasant in the idea.

Geese live and breed to a great age; in some cases, it is said, till at least forty years old!

The Toulouse goose is not a good sitter, and it is best to set her eggs under a large, heavy hen.

There are many other beautiful and curious breeds of geese, of which the beautiful:

Canada Goose

This is not uncommon on the ornamental waters of our English parks.

The Sebastopol Goose

This is really a sort of White Frizzled Goose, the plumage being not only turned back like Frizzled Fowls, but very silky and soft in texture.

The Chinese Goose

Of two principal colours:

1. The Brown, with dark brown or black bill and brown eyes, and a darker brown stripe down the neck and back, and dark orange shanks.
2. The White, with orange bill, deep blue eyes, and orange shanks. The *Book of Poultry* says the White Chinese should also have a brown band down the back, but the American Standard says the plumage should be pure white. They have a large brown or black knot at base of bill.

Index